PRACTICAL DATA ANALYSIS

CASE STUDIES IN BUSINESS STATISTICS
VOLUME I

PRACTICAL DATA ANALYSIS

CASE STUDIES IN BUSINESS STATISTICS
VOLUME I

SECOND EDITION

Peter G. Bryant
Marlene A. Smith
College of Business and Administration
and
Graduate School of Business Administration
University of Colorado at Denver

Boston   Burr Ridge, IL   Dubuque, IA   Madison, WI   New York   San Francisco   St. Louis
Bangkok   Bogotá   Caracas   Lisbon   London   Madrid
Mexico City   Milan   New Delhi   Seoul   Singapore   Sydney   Taipei   Toronto

Irwin/McGraw-Hill

A Division of The McGraw·Hill Companies

PRACTICAL DATA ANALYSIS: CASE STUDIES IN BUSINESS STATISTICS VOLUME I

Copyright ©1999 by The McGraw-Hill Companies, Inc. All rights reserved. Previous edition ©1995 by Richard D. Irwin, Inc. Printed in the United States of America. Except as permitted under the United States Copyright Act of 1976, no part of this publication may be reproduced or distributed in any form or by any means, or stored in a data base or retrieval system, without the prior written permission of the publisher.

This book is printed on acid-free paper.

2 3 4 5 6 7 8 9 0 DOC/DOC 9 3 2 1 0 9

ISBN 0-256-23871-5

Vice president/Editor-in-chief: *Michael W. Junior*
Publisher: *Jeffrey J. Shelstad*
Senior sponsoring editor: *Scott Isenberg*
Developmental editor: *Wanda J. Zeman*
Marketing manager: *Zina Craft*
Senior project manager: *Susan Trentacosti*
Senior production supervisor: *Heather D. Burbridge*
Cover designer: *Steven Vena/SrV Unlimited Design*
Compositor: *Carlisle Communications, Ltd.*
Typeface: *10/12 Times Roman*
Printer: *R. R. Donnelley & Sons Company*

Library of Congress Cataloging-in-Publication Data

Bryant, Peter G.
    Practical data analysis : case studies in business statistics /
Peter G. Bryant and Marlene A. Smith.—2nd ed.
      p. cm.
    Includes bibliographical references (p.).
    ISBN 0-256-23871-5 (vol. 1). —ISBN 0-256-23872-3 (vol. 2). —
ISBN 0-07-365488-4 (vol. 3)
    1. Commercial statistics—Case studies. 2. Commercial statistics-
–Data processing. 3. Statistics—Data processing. I. Smith,
Marlene A. II. Title.
HF1017.B79   1999
650'.01'5195—dc21   98-43159

http://www.mhhe.com

To L.S.B. and L.G.B.

In memory of Reuben and Charlotte Aschenbach

And he who is versed in the science of numbers can tell of the regions of weight and measure, but he cannot conduct you thither.

Kahlil Gibran
The Prophet

Was man nicht nützt, ist eine schwere Last.
Johann Wolfgang Goethe
Faust I, 684

Some of the blocks to dealing comfortably with numbers . . . are due to quite natural psychological responses to uncertainty, to coincidence, or to how a problem is framed. Others can be attributed to . . . romantic misconceptions about the nature and importance of mathematics.

John Allen Paulos
Innumeracy

INTRODUCTION FOR THE STUDENT

INTRODUCTION

This book contains a collection of *cases* or situations. Most cases contain some *data*. In others, we ask you to gather data or figure out how you *would* gather the data. Presumably the data can shed some light on the situation. Your job is to decide what light the data shed. To do this, you will have to

- Understand the situation and context.
- Choose an appropriate technique to summarize or analyze the data.
- Decide what you think the data have to say.

Then you will have to communicate your findings, in a way that's effective in the context of the situation described, whether that be a presentation to a board of directors, a memo to your boss, or something else.

CASES

In some fields of education—law or medicine, for example—you often "solve" a case by applying your knowledge to the situation given in a systematic way, and if you apply your knowledge correctly, you arrive at a single correct, well-defined answer, like a legal opinion or a diagnosis. Different people applying the course material correctly will arrive at the same answer.

In *business* education, cases usually play a different role: they are vehicles to raise issues, point out inherent conflicts, and stimulate debate. At the end of the debate, you may have settled some issues, but others remain open. Your answers or conclusions will not necessarily agree with those of your colleagues, for they will have brought different assumptions, interpretations, and solution methods to the debate.

We expect that you will *debate* these cases. We designed them that way. You may debate various aspects of the situation:

- The analysis method to use.
- The quality of the data.
- The proper interpretation of the analysis results.
- The recommendations or conclusions to draw.

For example, you may find you agree with your colleagues about the appropriate analysis of the data, but disagree about whether the data are of a quality that justifies a conclusion. Sometimes the analyses will be easy, sometimes hard. Sometimes a single method of attack seems obvious; at other times the situation yields to various different approaches. Some cases seem to require relatively sophisticated methods for full solution, but simple techniques will lead to at least part of the message. All the cases can profit from discussion and debate, though; they are not purely statistics or computational homework exercises.

REMARKS

Your instructor will specify the formats he or she wants you to use for your presentation and written summary, and will guide any debate. We often ask students to prepare a one-page summary (perhaps augmented by a chart or graph) and to come to class prepared to present and/or defend their findings in a general class discussion. This works particularly well when you prepare your report in a group. You will find it takes a lot of work to get a group of three to five people to agree on what the message in the data is! When you have to boil it down to one page, it puts even more pressure on you to know exactly why you did what you did and what you think it means. Often, other students will use different techniques and/or come to different conclusions, and you will have to defend your approach and conclusions to them. That's part of the fun and the frustration: statistics is not meant to be a sterile exercise at the end of a textbook chapter. It's supposed to be techniques for real people to better inform themselves about real situations by using real data. That sometimes means ambiguity, controversy, assumptions, disagreement and hard work. We think you'll find it helpful, worthwhile, and relevant, though. That's the message from our students who've tried it this way.

There's no one "right" approach to any of these cases, although often we had a particular approach in mind when we wrote it. Almost every time we assign one of them, though, some student points out a valid alternative we hadn't thought about. And just because the case is in the book doesn't mean we think the results will be overwhelming. Sometimes, the data don't really have much to say. That happens in real life, and it may happen here, too.

The cases are given in random order, so you can't tell what technique we had in mind by where the case is in the book or by how you approached the case next to it.

Most of the cases include computer files with data, and most of the time you will want to use a computer but not always. The presence of a computer file doesn't necessarily mean we think that you should use a computer. For those instances in which you would *like* to use a computer, we have prepared data files in a variety of formats. Your instructor will give you more information about acquiring and using the data files.

In our classes, we place a lot of emphasis on how well students write up or present their findings. In practice, a good analysis may influence nobody if it isn't presented well, and we think you should make it a habit, right from the start, to put as much energy into your presentation as you do into the analysis. We don't particularly advocate fancy, color graphics or other "fashionable" approaches. Good hand-drawn graphs can be wonderfully effective, too. The key is to communicate clearly. We discuss report writing more in the next section.

We hope you have fun with the cases. If you come up with an example from your own life or career that you think others would enjoy or profit from, please let us know. Maybe the next edition will have room for it!

Peter G. Bryant
Marlene A. Smith

ACKNOWLEDGMENTS

Many colleagues have reviewed individual cases and the manuscript of *Practical Data Analysis*. We thank specifically:

Sung K. Ahn, *Washington State University*
Randy Anderson, *California State University at Fresno*
Bridget G. Heidemann, *Seattle University*
Raj Jagannathan, *University of Iowa*
Jerzy J. Letkowski, *Western New England College*
Walter J. Mayer, *University of Mississippi*
Dale McFarlane, *Oregon State University*
J. B. Orris, *Butler University*
Paul Paschke, *Oregon State University*
John R. Pickett, *Georgia Southern University*
Kipling M. Pirkle, *Washington and Lee University*
Al Schainblatt, *San Francisco State University*
Milo Schield, *Augsburg College*
William L. Seaver, *University of Tennessee*
Herbert Spirer, *University of Connecticut*
Jeffrey W. Steagall, *University of North Florida*
Stanley A. Taylor, *California State University at Sacramento*
Betty M. Thorne, *Stetson University*
Thomas K. Tiemann, *Elon College*
Wayne Winston, *Indiana University at Bloomington*

We acknowledge also the founders, organizers and participants in the annual conference entitled *Making Statistics More Effective in Schools of Business* for their hard work and interest in changing the way in which we teach statistics in our business classrooms.

For specific permissions we thank the American Statistical Association for permission to reproduce items from *The American Statistician;* the *Rocky Mountain News* for copyright permission; and Stanley D. Elofson for assisting us in obtaining copyright permission for the 1992 Arapahoe County property assessment study. Further acknowledgments are listed with individual cases.

Several friends, colleagues, and relatives helped in various ways. Edward J. Conry advised us on the preparation of our prospectus. Lucinda Bryant, Louise Bryant, and Lynwood Bryant read the book and commented on content, style, and presentation. Kathryn E. Chmelir and Adriano Pimenta helped us with research, data input and analysis, and other related tasks. The staff at Irwin/McGraw-Hill, including Scott Isenberg and Wanda Zeman, have also been helpful.

Finally, we thank our students and other contributors to the cases in the book. Where appropriate, we acknowledge them individually in the source notes of their respective cases. All of our students contributed indirectly to this project: they told us what did and didn't work, complained about bad cases (and rightly so!), and had fun with the good ones.

Peter G. Bryant
Marlene A. Smith

CONTENTS

PREPARING WRITTEN CASES AND BUSINESS REPORTS		I–xv
THE DATA DISK		I–xxi
CASE 1	Waste Paper	I–1
CASE 2	Dome Loading	I–5
CASE 3	Spot Prices	I–9
CASE 4	Joanne's Forecasting Problem	I–11
CASE 5	T-Shirt Designs	I–13
CASE 6	Stocks A and B	I–17
CASE 7	Tom's Used Mustangs	I–19
CASE 8	American History Illustrated	I–23
CASE 9	Healthy Lifestyles	I–25
CASE 10	Lake County Lunches	I–29
CASE 11	Machine Production Records	I–31
CASE 12	Kerrville Rainfall	I–35
CASE 13	Workload Ratings	I–37
CASE 14	Roaring Fork Transit	I–41
CASE 15	Sturgel Division	I–43

CASE 16	Photocopying Abuses	I–45
CASE 17	AgriComp	I–47
CASE 18	Tips	I–51
CASE 19	Rubbergate	I–55
CASE 20	HiTech Engineering	I–61
CASE 21	401(k)	I–65
CASE 22	Late Charts	I–69
CASE 23	Money Supply and Interest Rates	I–71
CASE 24	SAT Scores	I–79
CASE 25	Emergency Admissions	I–83

PREPARING WRITTEN CASES AND BUSINESS REPORTS

The cases in this book require you to communicate your statistical results somehow. You may discuss your results in class with your instructor and your other classmates. You may give formal classroom presentations. Often, you will prepare a written report. Writing and speaking effectively are important to your professional career. Here are some thoughts on writing.

There are many ways to organize a business memo, but formal structure is probably less important than clear, concise communication. Writing the summary of your results will probably be just as difficult as doing the computations. Our students say it takes about as long to write the results as it does to finish the statistical work. You might want to plan accordingly.

WRITING STYLE

Once you are done with the computer work, you must interpret these findings for someone, perhaps a manager or your boss.

Avoid Jargon

Write so that *anyone* can understand it. Don't use technical language to intimidate or overwhelm the recipient. Unless you are writing to someone well-versed in statistical techniques, don't use statistical and mathematical jargon in your report. If you were the vice president of marketing, and had never studied statistics, which of the following two summaries would you find most valuable?

- The regression model is given by: SALES = 13,000 + 5*ADVERTISING, which means that the y-intercept is $13,000 and the slope is $5.

- According to the model: (*a*) if the firm does no advertising, sales will be $13,000, and (*b*) each additional dollar spent on advertising will bring in five additional dollars in sales.

Your instructor and the textbook will use jargon. You will see specific statistical terms like "least squares" and "p-values" in class, but don't presume that the *reader* of your memo understands them. You must translate statistical concepts, methods, and outcomes for the uninitiated. By virtue of spending time in your statistics class, you may forget that certain statistical concepts don't exist in most people's vocabularies. Here are some guidelines.

- Don't use words or phrases in your report unless you used them before signing up for statistics.
- Would your next door neighbor understand the essence of your report?
- What would the editor of your local newspaper think about publishing your report in tomorrow's newspaper?

Use the Active Voice

Use the active voice. "I ate my dinner" says more than "The dinner was eaten by me," and says it better.

Be Clear

Say what you mean, unambiguously. "Nothing is capable of being well set to music that is not nonsense,"[1] is a quotable epigram, but does it mean "Nothing is capable of being well set to music-that-is-not-nonsense"? Or does it mean "Nothing-that-is-not-nonsense is capable of being well set to music"? If you mean "only nonsense can be set to music," say so.
Similarly,

"My face looks like a wedding cake left out in the rain."[2]

is great poetry, but leaves room for various interpretations.

ORGANIZATION

Introductory Material

To help your reader, provide some background information, such as a statement of the problem or situation and a description of the data available to answer this question. Since your report involves statistical analyses, you might also provide initial descriptive statistics (e.g., means and/or standard deviations) of the more important variables, unless that information would simply distract from your point. Such background information puts the problem in perspective and eases the reader into the upcoming material.

---

[1] Joseph Addison (1711), quoted in *The Columbia Dictionary of Quotations*. Copyright © 1993 by Columbia University Press. All rights reserved.

[2] W. H. Auden (1907–73), quoted in Humphrey Carpenter, *W. H. Auden*, pt. 2, ch. 6 (1981). *The Columbia Dictionary of Quotations*. Copyright © 1993 by Columbia University Press. All rights reserved.

Good Grammar

An effective report uses good grammar, correct spelling and appropriate punctuation. You can't convince a reader that your statistical results are valid if your writing is poor. Sloppy, misspelled, or disorganized reports send a message: you don't think the report is very important. Brush up on your writing skills. It's fun and valuable to write effectively. Word processors may help. A spelling checker is useful, too.

Junked-Up Appendixes

Consider putting supporting statistical documentation, such as graphs, tables, and other statistical output, at the end of your written report. Within the report, refer to these appendixes (e.g., "See Exhibit II") for guidance. On the other hand, if one particular table or graph contains the essence of the point you're making, put it with the text, where your reader can see it quickly. Ask yourself: If I put it in with the text, will it distract the reader more than if the reader has to turn to an appendix? A critical graph belongs in with the text. A table giving relatively minor support to your argument belongs in an appendix.

Don't append every statistical printout you produced. Include the *important* ones. If *you* can't decide which is important and which is not, how can your readers? In fact, if you haven't decided which are important, you're not done with your analysis yet. "Don't append a statistical exhibit if you don't refer to it in the report, and, of course, don't refer to an exhibit that's not there" might be a good guideline.

Length

We generally limit reports to one or two pages. Your instructor may impose different limits, but write as succinctly as you can. Present the essence of your statistical findings, not a comprehensive validation of every step of your work. Managers won't wade through pages and pages of information. It's easy to overwhelm a reader who wants nothing more than a summary of the major findings, but why do so?

You must identify the fine line between too much detail and not enough detail. Don't make your report so short that it's deceptive, but don't bore or intimidate your intended reader with unnecessary details, either. You, the reporter of the data, must decide what to include.

Even if your report is a lengthy project, we suggest that you make the first page an "executive summary," written for someone who has never had a statistics class, and doesn't care to have statistics explained. Include only the *important* results and implications derived from the data, and any necessary caveats or limitations of your findings. For instance, you might not trust your results because of a small sample size, or some confidence interval might suggest inaccuracy of a point estimate. Try to limit this executive summary to one page, single-spaced. It's hard to do!

To start you off, we've included sample reports on the next three pages. Preparing reports will become increasingly straightforward with practice. Good luck!

A PRETTY LOUSY BUSINESS REPORT

TO: Tom Jones, Director of Personnel

FROM: Jana Smith, Statistician

DATE: 5/19

SUBJECT: Analysis of Personnel Data

Per your request, I have analyzed the data on sick days and the company's new wellness program. The results are summarized below.

The regression model suggests that there is a statistically significant relationship between the two variables. The correlation between dollars contributed to the wellness program and absenteeism is a positive 0.63. The standard error is 0.073. The R-squared statistic on the simple regression model is 56%, which is pretty good for cross-sectional data. Moreover, the F-statistic measuring the statistical significance of the model as a whole is 54.90, indicating a good model.

When analyzing the relationship by gender, there is no statistically significant impact here. The p-value on the categorical variable called GENDER (see EXHIBIT II) is .46. This means that absenteeism does not seem to be correlated with the gender of the employee. On the other hand, it just might be a multicollinearity problem with the two independent variables.

I'd be happy to assist in any further analysis of this data at your request.

A PRETTY GOOD BUSINESS REPORT

TO: Tom Jones, Director of Personnel

FROM: Jana Smith, Statistician

DATE: 5/19

SUBJECT: Analysis of Personnel Data

As you asked, I have analyzed the data on sick days and the company's new wellness program. Here are my results.

Data

The personnel department provided a sample of 125 randomly chosen employee files. From those files, we obtained:

- the company's payment for that employee's participation in our wellness program,

- that employee's absentee record (measured in number of absent days) over the past two years, and

- the gender of the employee.

72% of the sample group was female.

Results

- On average, our employees missed 15 days of work in the first year. The typical fluctuation around the average of 15 days was seven days.

- In the second year, after starting the wellness program, the average number of absent days declined to 10 days, and the typical fluctuation also decreased to 2 days. We committed about $50 per employee to the wellness program last year.

- A statistical model (see EXHIBIT below) of the relationship between dollars committed to wellness, absenteeism, and gender indicates that the wellness program has a statistically significant relationship to absenteeism. The model suggests that:

    - each additional dollar committed to our wellness program was associated with a two hour decline in absenteeism, and

    - there is no statistically significant relationship between gender and absenteeism.

The model would generally be considered a statistically strong one, since 56% of the variation in absenteeism is explained by the model. Although this leaves 44% of the variation unexplained, it is difficult to do much better with the type of data available for this analysis.

Recommendation

Because of the decline in absenteeism after the institution of the wellness program, I recommend that the program be continued.

Limitations

Consider redoing this study in another year with a larger sample size. It is not clear whether one year is enough to observe the full benefits of the wellness program. Also, there are some troubling aspects of the statistical results that might be alleviated with a larger sample size. For instance, many of the standard errors in the model (that is, measures of the accuracy of the estimates) are quite large in my opinion.

I'd be happy to answer any further questions that you might have about my analysis or report.

EXHIBIT

```
The regression equation is
ABSENT = 12.2 - 0.26 DOLLARS + 0.07 GENDER

Predictor         Coef        Stdev       t-ratio        p
Constant          12.2        3.37        3.62           0.00
DOLLARS          -0.26        0.04        6.50           0.00
GENDER            0.07        0.25        0.28           0.80

s = 7.9     R-sq = 56.2%   R-sq(adj) = 55.5%

Analysis of Variance

SOURCE          DF          SS          MS          F          p
Regression       2         9770        4885        78.3       0.000
Error          122         7614        62.4
Total          124        17384
```

THE DATA DISK

Most of the cases in *Practical Data Analysis* require you to analyze a data set. A printed version (hard copy) of the data set may be found at the end of each case. The printed version lets you get a quick, informal sense of the data. At some point, you will probably want to use a statistical or spreadsheet software package for more formal analyses of these numbers. Some data sets are small enough to enter the data yourself. For others, that task would be very time consuming.

Your instructor has access to the data files for the cases in *Practical Data Analysis*. She or he will distribute those files to you in a form that is appropriate for the computing environment at your school. You may also receive instructions on how to copy the files for your own use and storage, and how to feed the data into one or more software programs.

Once you receive the data in machine-readable form from your instructor, you should be able to get up and running quickly.

PRACTICAL DATA ANALYSIS

CASE STUDIES IN BUSINESS STATISTICS
VOLUME I

CASE 1

WASTE PAPER

Waste accompanies any production process. In this case, the production process is a printing press, which produces small books for a variety of customers. Data from 127 runs of this press, gathered over about a month, are listed on the next pages and are available in file WASTEPAP.

In most press runs, some books are wasted, and the number of wasted books (WASTE) is influenced by such factors as NET (the number of books requested by the customer) and GROSS (the total number of books actually produced to satisfy the customer's request. GROSS = WASTE + NET). Moreover, some books are more complicated to produce than others. Books of 8 or 16 pages (PAGES) can be produced using a single "web" of paper, while books of 24 or 32 pages require two webs, and this increases the chance of some sort of error in the final book. Similarly, books with larger numbers of colors (COLORS) may be more prone to errors, as the separate colors must be properly aligned with each other to produce a satisfactory product. The company's usual book is four-color, but occasionally other variations occur. The basis weight of the paper used (PAPER, in pounds) may also influence waste.

In pricing its four-color services, the company assumes a common "waste factor" for all of the different sizes of book (8-paged, 16-paged, etc.). Price quotes for other than four-color books are handled separately, but the analyst wants to know whether the "waste factor" really should be the same for all sizes of four-color books. Can you advise her?

The case is based on a real situation, described to us by Don W. Hansen, V.P. Operations, American Web, Inc., who also provided the data.

CASE 1 WASTE PAPER

DATA SET

OBS	NET	GROSS	WASTE	PAGES	PAPER	COLORS
1	52700	64700	12000	24	70	1
2	4800	7700	2900	8	50	2
3	13100	16700	3600	16	60	4
4	72000	77400	5400	16	50	4
5	72000	77200	5200	16	50	4
6	72000	80100	8100	16	50	4
7	26700	33100	6400	8	80	4
8	26700	31900	5200	32	60	4
9	29100	37100	8000	24	70	1
10	10300	14700	4400	32	50	1
11	10300	14000	3700	16	50	4
12	10300	15600	5300	16	70	4
13	7900	11500	3600	32	70	4
14	7900	10600	2700	16	60	4
15	20800	23700	2900	16	40	1
16	16300	19900	3600	8	40	4
17	17900	20200	2300	8	40	4
18	53100	58600	5500	16	47	4
19	53100	60000	6900	16	47	4
20	26200	37300	11100	16	60	4
21	26200	34400	8200	32	60	4
22	37300	45800	8500	32	60	4
23	52500	57500	5000	32	60	1
24	23500	28300	4800	32	60	1
25	23690	31000	7310	8	40	4
26	17900	24000	6100	16	40	4
27	28325	34400	6075	32	47	4
28	28325	33500	5175	32	47	4
29	28325	33000	4675	16	47	4
30	50000	59400	9400	16	47	4
31	51200	57900	6700	16	40	5
32	51200	57900	6700	16	40	4
33	11000	14600	3600	16	40	4
34	7800	10800	3000	16	60	4
35	3200	6300	3100	16	60	4
36	20000	25300	5300	16	70	4
37	26550	32200	5650	16	60	4
38	26550	32200	5650	16	40	5
39	26550	36300	9750	24	40	4
40	19570	24700	5130	32	70	4
41	16000	18100	2100	16	70	2
42	24383	34000	9617	32	70	4
43	24383	30600	6217	32	50	4
44	36300	41800	5500	32	50	4
45	36380	42900	6520	32	50	4
46	14200	18200	4000	16	50	4

(continues on the following page)

OBS	NET	GROSS	WASTE	PAGES	PAPER	COLORS
47	14200	17200	3000	8	50	4
48	10800	13900	3100	32	50	1
49	11900	14800	2900	16	50	4
50	10800	15400	4600	32	70	4
51	31300	38000	6700	24	40	4
52	52700	62300	9600	16	70	6
53	13100	16300	3200	16	60	4
54	4350	9200	4850	32	50	4
55	66560	82000	15440	16	60	5
56	66560	82100	15540	24	60	2
57	66560	83300	16740	32	60	2
58	36000	46600	10600	8	100	4
59	72000	83700	11700	16	50	4
60	43000	48400	5400	16	50	4
61	72000	78500	6500	16	50	4
62	72000	81400	9400	16	50	5
63	11300	14500	3200	16	80	4
64	5700	9800	4100	8	80	4
65	29000	34500	5500	16	70	4
66	29000	35400	6400	32	47	4
67	29000	35700	6700	32	47	1
68	10300	17000	6700	24	60	1
69	10300	19700	9400	32	60	4
70	17300	27900	10600	32	50	1
71	17300	22500	5200	32	70	4
72	14000	21900	7900	24	60	4
73	14000	20800	6800	16	70	4
74	15600	19100	3500	8	60	1
75	20900	28900	8000	24	60	4
76	15600	20400	4800	24	60	4
77	15600	24400	8800	32	40	4
78	11300	16500	5200	8	80	4
79	5150	11200	6050	8	100	4
80	53100	61900	8800	16	47	4
81	53100	58600	5500	16	47	4
82	53100	57300	4200	16	47	4
83	26200	34700	8500	32	60	1
84	8659	14900	6241	16	50	4
85	15028	18700	3672	8	47	4
86	52500	55100	2600	8	60	1
87	52500	57400	4900	16	60	4
88	52500	56800	4300	16	60	4
89	52500	58900	6400	16	60	4
90	23690	26000	2310	16	40	4
91	41000	45200	4200	8	40	1
92	41000	54100	13100	32	47	4
93	17100	22300	5200	32	60	4
94	12650	16600	3950	16	70	4

(concludes on the following page)

CASE 1 WASTE PAPER

OBS	NET	GROSS	WASTE	PAGES	PAPER	COLORS
95	12650	16600	3950	24	47	4
96	14162	19500	5338	8	100	4
97	14400	18500	4100	16	45	4
98	14400	19200	4800	16	45	4
99	14400	20700	6300	16	60	5
100	20850	26200	5350	32	70	4
101	28325	35200	6875	16	47	4
102	11400	14200	2800	16	40	4
103	5625	10000	4375	16	50	4
104	5625	8200	2575	16	50	4
105	5625	11300	5675	16	50	4
106	50000	56800	6800	16	47	4
107	50000	60300	10300	16	47	4
108	50000	56000	6000	16	47	4
109	50000	56000	6000	16	47	4
110	51200	61000	9800	16	70	4
111	38100	48000	9900	24	50	2
112	38700	51100	12400	32	60	4
113	54075	62400	8325	24	45	4
114	54075	63000	8925	32	45	2
115	27037	31100	4063	32	100	4
116	20000	35000	15000	32	47	4
117	27150	37400	10250	24	40	4
118	20000	32000	12000	32	47	4
119	18050	26000	7950	8	60	4
120	36380	45200	8820	16	50	4
121	7200	11100	3900	8	70	4
122	36380	40500	4120	16	50	4
123	36380	42000	5620	32	50	4
124	36380	41700	5320	32	50	4
125	24000	35500	11500	32	60	4
126	15450	21100	5650	16	60	4
127	15450	23300	7850	16	80	4

CASE 2

DOME LOADING

Circle Electronics manufactures keyboards and data-entry devices. Most of the company's products contain one or more bubble-like steel parts called domes. The domes are placed on a mylar surface. When a key on the keyboard is pressed, it physically pushes down on a dome and that causes appropriate electrical contacts. The computer receives electronic signals (corresponding to the key pressed) as a result.

A robot assembly called (naturally) a *dome loader* loads the domes onto the mylar surface. The process works as follows. The mylar sheet comes in 20" × 27" sheets. These sheets are fed manually into the dome loader, which takes domes one at a time from a reel and places them on the mylar sheet, at locations specified by X and Y specifications. The X and Y specifications give the horizontal and vertical distances, respectively, which specify the desired location of the center of the circular dome. The diagram on the next page gives two domes. Dome A has X and Y specifications of 1.425 and .800 inches, respectively, while Dome B has X and Y specifications of 2.300 and 1.625 inches, respectively. Once the domes have been placed on the mylar sheet, the sheet is fed manually into a machine that cuts it to its final size by using a steel rule die-cutting process.

If the resulting keyboard or similar product is to work reliably, the process must produce a sheet with accurately placed domes. The industry currently considers dome placement "acceptable" if it is within plus or minus .01 inches of the X and Y specifications, a range that is generally attainable by hand tools.

Circle Electronics studied one product to see if the dome-loading process was producing acceptable sheets and to provide any information that seemed helpful to the manufacturer of the dome loader in correcting any problems. The X and Y specifications for the domes are listed in the data that follow. Five consecutive sheets were measured. Each

---

This case is based on a real industrial situation, though the name of the company has been disguised. Brad A. Strong, Vice President, Operator Interface Technology, provided the data.

sheet contained nine domes. The data include the following variables (available in file DOMELOAD). SHEET is the sheet number and DOME is the dome number (the dome loader first places Dome 1 on the mylar sheet, then Dome 2, etc.). XMEAS is the actual X distance to the center of the dome; XSPEC is the X specification for this dome. YMEAS is the actual Y distance to the center of the dome; YSPEC is the Y specification for this dome.

The XMEAS and YMEAS variables were measured after the completion of the manufacturing process by special, accurate measurements. The XSPEC and YSPEC variables are the same for corresponding dome numbers and come from the manufacturing specifications for the particular product.

Is the process behaving acceptably? Are there any patterns that suggest how variation in the process could be reduced?

DIAGRAM

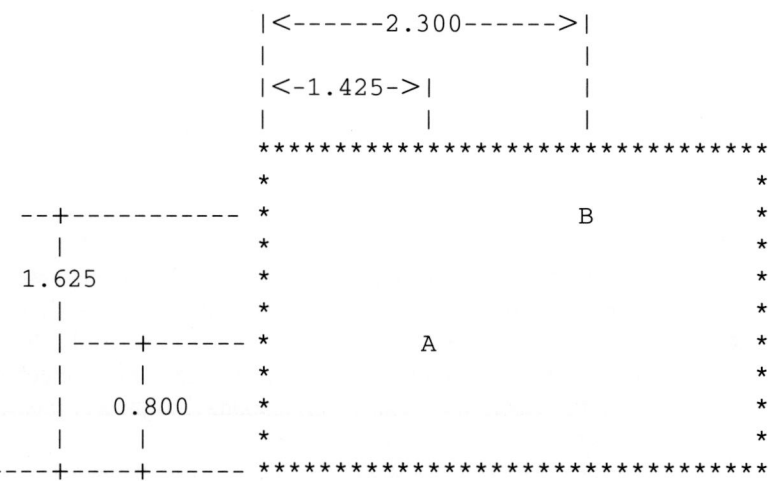

DATA SET

OBS	SHEET	DOME	XMEAS	XSPEC	YMEAS	YSPEC
1	1	1	1.439	1.425	0.804	0.800
2	1	2	2.321	2.300	1.625	1.625
3	1	3	3.853	3.863	2.262	2.250
4	1	4	5.120	5.113	2.253	2.250
5	1	5	2.928	2.925	2.885	2.875
6	1	6	5.439	5.425	2.896	2.875
7	1	7	7.311	7.300	2.888	2.875
8	1	8	7.299	7.300	1.641	1.625
9	1	9	7.859	7.838	0.195	0.175
10	2	1	1.416	1.425	0.805	0.800
11	2	2	2.285	2.300	1.631	1.625
12	2	3	3.860	3.863	2.258	2.250
13	2	4	5.098	5.113	2.255	2.250
14	2	5	2.916	2.925	2.900	2.875

(concludes on the following page)

OBS	SHEET	DOME	XMEAS	XSPEC	YMEAS	YSPEC
15	2	6	5.413	5.425	2.884	2.875
16	2	7	7.288	7.300	2.884	2.875
17	2	8	7.296	7.300	1.655	1.625
18	2	9	7.826	7.838	0.185	0.175
19	3	1	1.430	1.425	0.805	0.800
20	3	2	2.301	2.300	1.612	1.625
21	3	3	3.867	3.863	2.237	2.250
22	3	4	5.116	5.113	2.254	2.250
23	3	5	2.920	2.925	2.883	2.875
24	3	6	5.432	5.425	2.868	2.875
25	3	7	7.296	7.300	2.868	2.875
26	3	8	7.299	7.300	1.632	1.625
27	3	9	7.833	7.838	0.178	0.175
28	4	1	1.412	1.425	0.793	0.800
29	4	2	2.292	2.300	1.623	1.625
30	4	3	3.860	3.863	2.248	2.250
31	4	4	5.102	5.113	2.252	2.250
32	4	5	2.931	2.925	2.892	2.875
33	4	6	5.425	5.425	2.881	2.875
34	4	7	7.307	7.300	2.883	2.875
35	4	8	7.295	7.300	1.631	1.625
36	4	9	7.828	7.838	0.181	0.175
37	5	1	1.412	1.425	0.793	0.800
38	5	2	2.292	2.300	1.611	1.625
39	5	3	3.852	3.863	2.238	2.250
40	5	4	5.102	5.113	2.236	2.250
41	5	5	2.919	2.925	2.873	2.875
42	5	6	5.413	5.425	2.859	2.875
43	5	7	7.294	7.300	2.862	2.875
44	5	8	7.294	7.300	1.619	1.625
45	5	9	7.834	7.838	0.172	0.175

CASE 3

SPOT PRICES

All the way home from class, Sarah could hardly concentrate on the traffic. She was so excited about the prospect of her future wealth that she almost got into several accidents.

All this enthusiasm was sparked by Sarah's finance class at the University of Oklahoma. The lecture for the day was about spot markets. Her ears really picked up during the lecture and, even before she was out the door, Sarah had discovered a scheme that could allow her to make money on the spot market for natural gas.

The spot market for natural gas is that market in which natural gas is sold under a contract term of no more than one month. Having worked for some time for an Oklahoma pipeline company, Sarah knows a couple of important things about the spot market for natural gas. First, she knows that the spot price for Louisiana natural gas is generally known on or before the first day of trading on the Oklahoma market. Second, she knows that the Oklahoma spot price typically runs about 85% of the Louisiana spot price, even throughout the seasonal variations in these two markets. Therefore, in Sarah's mind, it should be easy to predict the current Oklahoma spot price and buy or sell futures based on the current Louisiana spot price. This would allow her to pay off her student loans and go to Jamaica for spring break.

The data set, in the file SPOTS, contains 46 monthly observations on Louisiana and Oklahoma spot prices for natural gas. In this file, MONTH represents the appropriate month (with one equaling January and so forth) of that YEAR. OSPOT is the Oklahoma spot price for natural gas measured in dollars per million British thermal units ($/MMBtu), and LSPOT is the Louisiana spot price, also in $/MMBtu. What formula will allow Sarah to play the spot market successfully? What is Sarah's forecast for the Oklahoma spot price for November 1997 if the November 1997 Louisiana spot price was just announced to be 1.89 $/MMBtu?

CASE 3 SPOT PRICES

DATA SET

MONTH	YEAR	OSPOT	LSPOT
1	94	1.985	2.215
2	94	2.008	2.138
3	94	1.753	1.968
4	94	1.442	1.643
5	94	1.372	1.533
6	94	1.350	1.528
7	94	1.375	1.583
8	94	1.420	1.693
9	94	1.577	1.863
10	94	1.610	1.838
11	94	1.743	2.058
12	94	1.890	2.357
1	95	1.913	2.323
2	95	1.582	1.890
3	95	1.357	1.608
4	95	1.383	1.665
5	95	1.483	1.792
6	95	1.518	1.837
7	95	1.488	1.793
8	95	1.485	1.725
9	95	1.418	1.663
10	95	1.425	1.705
11	95	1.557	1.915
12	95	1.860	2.252
1	96	2.313	2.570
2	96	1.733	2.132
3	96	1.373	1.637
4	96	1.362	1.618
5	96	1.362	1.600
6	96	1.387	1.610
7	96	1.362	1.568
8	96	1.330	1.490
9	96	1.350	1.503
10	96	1.522	1.718
11	96	1.917	2.202
12	96	2.013	2.387
1	97	1.737	1.980
2	97	1.333	1.490
3	97	1.318	1.468
4	97	1.318	1.470
5	97	1.315	1.448
6	97	1.275	1.408
7	97	1.130	1.267
8	97	1.175	1.287
9	97	1.397	1.518
10	97	1.723	1.882

CASE 4

JOANNE'S FORECASTING PROBLEM

Joanne had written memos to the computer center director for months to complain about unsatisfactory turnaround time for the jobs her department ran on the computer. Basically, her department submitted regular jobs to process orders, write invoices, resupply parts, and/or change customer records. The same computer program processed all of these transactions, in whatever mix of orders, invoices, resupply requests, or customer change requests the department happened to request on that day. Joanne felt the computer center charges were outrageous, anyway. The combination of (what she perceived as) high rates and poor service frustrated her. It also caused tension in the various meetings that she and the computer center manager attended together.

Now, though, she was in a quandary. The computer center manager had just written to her, explaining that he was now prepared to provide whatever capacity was necessary to meet her turnaround time requirements, if she would simply forecast the capacity she needed, in terms of CPU minutes per month. After some inquiries, Joanne determined that CPU minutes were a measure of the computer time used for a job. The computer center manager's memo was a blessing and a curse: a blessing in that the computer center would respond to her needs, and a curse because while she was prepared to forecast her needs in her own units (she forecasted a monthly average of 8,000 orders, 6,500 invoices, 700 resupply requests, and 30 customer changes per month, based on the latest business plan), she had no real notion of how to convert them into CPU minutes, or any of the other kinds of units that the "computer types" seemed to talk about.

Glancing over her monthly bills from the computer center for the last year, she wondered if she could use that data to convert her forecast to the units the computer center

This case is an amalgam of several real situations and data sets from the authors' experience.

CASE 4 JOANNE'S FORECASTING PROBLEM

required. The relevant information from her computer center bills from the 12 months of last year is listed below. How should Joanne proceed? The data are also available in file JOANNE.

DATA SET

MONTH	ORDERS	INVOICE	RESUPPLY	CUSTOMER	CPUMIN
1	7000	4026	487	28	2281
2	6046	6073	619	46	2203
3	8031	4096	892	29	2602
4	4903	6023	900	11	1806
5	6208	4016	703	18	2072
6	4902	7773	402	47	2163
7	9003	4096	450	48	2689
8	9604	6032	706	57	3204
9	10618	10068	719	12	3921
10	10497	11063	850	9	4016
11	6033	6001	906	18	2168
12	2014	3004	402	18	902

CASE 5

T-SHIRT DESIGNS

Julio has one semester left in his MBA program and is looking at paying back some hefty student loans. Even though he has had to borrow money to complete the program, he's decided it was a worthwhile investment. Nonetheless, Julio would like to repay some of his loans before graduating if at all possible.

Julio has recently taken entrepreneurial and small-business classes and has decided to put some of this knowledge to work. He suspects that a number of his fellow students would be willing to shell out a few bucks for school memorabilia. Since summer is right around the corner, Julio is pretty sure that he can make some money by selling T-shirts to students on campus.

Informal conversations with people in the hallways have given him some ideas about popular T-shirt designs. Julio has narrowed down the possibilities to two. He's given sketches to a local T-shirt printing company, and they have made two prototype shirts. At this stage, Julio has no idea about the relative demand for the two T-shirt designs. If he's going to all the trouble and expense of having the shirts made up, it would be useful to know which of the two shirts is more popular.

The two shirts look something like this. The red shirt has a scoop-neck and is made out of 100% cotton. On the back of the shirt is a logo of the school. Under the logo are the words "PARTY TIME at PU UNIVERSITY!" Julio suspects that this shirt will mostly appeal to the party-goers.

The other shirt is more conservative. It is a white, button-down-collar shirt made of 50% cotton and 50% synthetic materials. This shirt has a simple, and small, version of the school logo on the upper-right-hand section of the front of the shirt.

Julio has decided to gauge the relative markets for his two T-shirts by running a quick survey. With permission from the student government, he has set up a table in the

This case was based on a survey designed by M.A.Smith, who also wrote the survey questionnaire.

student union. The two prototype shirts are there for inspection, along with a survey designed to gauge people's reactions to the two shirts. He has a big sign over the table offering a free soft drink to all participants to induce people to stop by the table and participate in the survey. On the corner of his table is a big cardboard box with a slit in the top. Students are instructed to put their completed questionnaires in the box, thereby guaranteeing anonymity of the responses.

The actual survey and the resulting data set are shown below. The raw data reside in the file named T-SHIRTS. Julio has hired you to analyze these data. Prepare a report about the preferences revealed by the data. Comment on whether the other survey variables relate somehow to preferences. As a final note, feel free to comment on the survey itself and give any suggestions that you might have for improving it.

SURVEY

```
I am trying to measure people's preferences for one of two
different T-shirt designs. Would you please help me with my
study by answering each of the following questions to the
best of your ability. Do not put your name on this page be-
cause all responses should be anonymous.

1. My gender is

      1 = female
      0 = male

2. My current or most recent scholastic grade point average
   on a 4-point scale is _____.

3. Circle the number that most accurately describes your at-
   titude about the red and white T-shirts on the table.

      1 = I strongly prefer the red T-shirt to the white
          T-shirt.
      2 = I somewhat prefer the red T-shirt to the white
          T-shirt.
      3 = I like both equally, or am indifferent between the
          two.
      4 = I somewhat prefer the white T-shirt to the red
          T-shirt.
      5 = I strongly prefer the white T-shirt to the red
          T-shirt.

4. I am _____ years old.

5. My gross annual income last year was $_____.
   (If you were gainfully unemployed, enter 0.)

6. I am (circle one)

      1 = left-handed
      2 = right-handed
      3 = ambidextrous (both left- and right-handed)
```

DATA SET

GENDER	GPA	PREFER	AGE	INCOME	HANDED	GENDER	GPA	PREFER	AGE	INCOME	HANDED
0	4.00	4	34	20.0	2	0	3.60	3	27	32.0	2
0	2.90	1	23	12.0	2	0	4.00	4	26	25.0	1
1	3.70	4	32	27.0	2	1	4.00	4	25	25.5	2
1	2.50	4	23	22.0	2	1	3.00	4	37	52.0	2
1	3.90	5	27	20.0	1	0	3.78	4	29	28.5	2
0	4.00	5	39	75.0	2	1	3.00	5	26	0.0	2
1	3.40	2	24	24.0	2	1	3.70	2	24	18.0	2
0	3.50	2	27	30.0	2	0	4.00	5	31	60.0	2
0	4.00	4	28	24.5	2	0	2.97	2	23	14.0	2
0	4.00	2	43	45.0	2	1	3.00	5	29	0.0	2
0	3.40	3	28	30.0	2	1	3.61	5	32	50.0	2
1	3.40	2	23	0.0	1	0	4.00	2	46	65.0	2
0	3.00	2	26	30.5	1	0	3.50	4	31	0.0	2
1	3.70	5	40	24.0	2	1	3.40	4	25	25.0	2
0	3.90	4	31	30.0	2	0	3.50	4	25	20.0	2
0	3.80	2	30	0.0	2	1	3.58	4	24	0.0	2
1	3.10	2	36	38.0	2	1	3.20	4	27	20.0	2
0	3.00	2	26	32.5	2	0	3.00	4	24	25.0	2
0	3.85	2	30	40.0	2	1	3.06	4	25	0.0	2
0	3.20	2	33	23.0	2	0	3.20	4	24	18.0	2
0	2.00	4	29	24.5	2	0	4.00	2	25	25.0	2
0	3.00	3	25	25.0	2	1	3.00	2	24	30.0	2
0	3.70	2	29	30.0	2	0	4.00	2	23	16.0	2
0	3.50	4	29	40.0	2	0	3.20	3	27	32.0	2
1	3.80	2	26	35.0	2	0	3.00	3	30	0.0	2
0	3.85	2	46	27.0	1	1	4.00	2	29	0.0	2
1	4.00	4	26	24.0	2	1	2.85	2	24	0.0	2
1	3.50	2	24	0.0	2	1	3.00	4	40	0.0	2
1	3.80	2	33	49.0	2	1	4.00	5	29	40.0	2
0	4.00	5	27	34.0	2	1	3.30	2	25	15.0	2
0	3.50	2	27	45.0	2	1	4.00	2	47	46.0	2
1	3.30	1	24	31.0	2	1	3.50	1	37	60.0	2

CASE 6

STOCKS A AND B

The data that follow (and in file STOCKAB) give the rates of return over a 50-year period for two stocks, which we'll call stock A and stock B. The rate of return is defined as the increase in value of the portfolio (including any dividends or other distributions) during the year divided by its value at the beginning of the year. That's the fraction by which your wealth would have changed had it been invested in that particular combination of securities. The rate of return may be either positive or negative.

What was the average return for stock A? Stock B? In the theory of finance, the standard deviation of the return is often used as a measure of the risk associated with investing in a given security. What are the standard deviations of the returns for stocks A and B?

Assume that the history of stocks A and B is a useful guide to what may be expected of them in the foreseeable future. (That may seem like a big assumption, but assume it anyway.) Suppose you want to use this history as a guide to making an investment decision for the long run. Based on this history, and assuming that standard deviation of return is the appropriate measure of risk, does it make any difference whether you invest wholly in stock A, wholly in stock B, or half in stock A and half in stock B? Assume that average return is to be maximized and risk is to be minimized, if that is jointly possible.

DATA SET

YEAR	A	B	YEAR	A	B	YEAR	A	B	YEAR	A	B
1	0.215	0.199	14	0.122	0.122	27	0.001	0.112	40	0.215	0.045
2	0.194	0.017	15	0.265	0.103	28	0.122	0.294	41	0.315	0.419
3	0.130	0.112	16	0.237	0.218	29	0.230	0.055	42	0.008	0.199
4	0.337	0.218	17	0.151	0.055	30	-0.013	0.179	43	0.058	0.055
5	0.165	0.246	18	0.201	0.294	31	0.044	0.132	44	-0.035	0.151
6	0.115	0.160	19	0.230	0.304	32	0.258	0.112	45	0.265	0.227
7	0.308	0.218	20	0.194	0.246	33	0.280	0.218	46	0.215	0.151
8	0.130	0.294	21	0.008	0.112	34	0.122	0.246	47	0.211	0.093
9	0.330	0.170	22	0.265	0.313	35	0.072	0.227	48	0.230	0.221
10	0.294	0.103	23	0.222	0.017	36	0.187	0.266	49	0.115	0.218
11	0.244	0.055	24	0.144	0.333	37	0.130	0.294	50	0.106	0.025
12	0.401	0.179	25	0.108	0.103	38	0.101	0.074			
13	0.251	0.428	26	0.194	0.122	39	0.308	0.246			

CASE 7

TOM'S USED MUSTANGS

Tom has recently opened a used car lot. Although he has very little business experience, his passion has always been for vintage automobiles. Tom has been collecting and repairing old cars since he was a teenager and has lots of technical knowledge about carburetors, shock absorbers, air conditioners, and so forth. Tom has decided to turn his hobby into a money-making venture for himself. Unfortunately, Tom knows almost nothing about advertising, taxes, social security payments for his employees, bookkeeping, or other aspects of the *business* of selling cars. For this reason, Tom is taking a few business courses at the local community college.

The car lot has been open for a few months now. Several of Tom's customers have told him that his asking prices are way out of line with the rest of the market—sometimes too high and sometimes too low, but never close to the going rate. Tom typically relies on a gut feeling for determining asking prices. For instance, all else the same, he thinks that red cars appear sportier and can be sold at a premium. He knows that older cars sell for less, but he doesn't know exactly how much less. Apparently this informal scheme is not working very well.

As an alternative, Tom has secretly visited a competitor's lot, but there are so many physical characteristics of a car that he is having trouble pinpointing a useful pricing scheme. Finding perfect matches for his cars is nearly impossible given all of the possible options on cars.

To more thoroughly understand this issue, Tom has hired you to perform a statistical analysis of asking prices for used Mustangs. Tom has already collected information from the local newspapers and other sources. The data file named MUSTANGS contains observations on 35 used Mustangs and 10 different characteristics.

This data set was assembled by Marla S. McClure and Larry Welner as part of a course project. Data collection involved surveying 35 private owners and dealers who advertised in various media on March 27, 1992.

CASE 7 TOM'S USED MUSTANGS

Prepare a report for Tom on the influence of various options on asking price, and show him how he should use this information to set prices on other cars in the near future.

DEFINITIONS

PRICE	Asking price in dollars
CONVERT	1 = convertible 0 = not convertible
AGE	Age of the car in years
MILES	Approximate odometer reading in miles
TRANS	1 = automatic transmission 0 = manual transmission
AIR	1 = air conditioner 0 = no air conditioner
CYL	Number of cylinders in the engine
COLOR	1 = maroon 2 = silver 3 = gray 4 = red 5 = blue 6 = black 7 = white
GT	1 = GT model 0 = not a GT model
OWNER	0 = private owner 1 = dealer owner

DATA SET

PRICE	CONVERT	AGE	MILES	TRANS	AIR	CYL	COLOR	GT	OWNER
12688	0	1	8200	0	1	8	1	1	1
11500	0	1	27000	1	1	8	2	0	0
11650	0	1	3000	0	1	8	3	1	1
10995	1	2	27000	1	1	8	4	0	1
6200	0	3	55000	1	1	4	4	0	0
9500	0	3	22000	0	0	8	3	0	0
8700	0	3	25000	1	1	8	5	0	0
12500	0	2	23800	0	1	8	4	1	1
10488	1	4	20000	0	1	8	7	1	1
5988	0	4	37000	0	1	4	3	0	1
7288	1	5	84000	1	1	4	5	0	1

(concluded on the next page)

CASE 7 TOM'S USED MUSTANGS

PRICE	CONVERT	AGE	MILES	TRANS	AIR	CYL	COLOR	GT	OWNER
3590	0	5	55000	0	0	4	6	0	0
2800	0	6	62000	0	1	4	5	0	0
6995	1	7	78000	1	1	4	1	0	0
4900	0	7	60000	0	1	8	7	1	0
2488	0	7	67000	1	1	4	1	0	1
4200	1	8	65000	1	1	6	5	0	0
12990	1	1	13300	1	1	8	7	0	1
14500	1	3	31000	1	1	8	7	1	0
10900	1	3	4000	1	1	8	6	0	1
13500	1	3	23000	0	1	8	6	1	0
12000	0	3	89000	0	1	8	7	1	1
8500	0	3	34000	1	1	8	7	0	0
4388	0	4	80000	0	1	4	5	0	1
3000	0	7	95000	0	1	4	5	0	0
3900	0	8	90000	0	1	8	1	1	0
3350	0	8	91000	0	1	8	6	1	0
10500	0	3	17000	0	1	8	1	1	0
12988	0	1	21000	1	1	8	7	1	1
6500	0	3	73000	0	1	8	7	0	0
12900	1	2	19000	0	1	8	6	0	0
9000	0	2	32000	0	1	8	5	0	0
7000	0	6	56000	0	1	8	7	1	0
5900	0	4	110000	1	1	8	5	0	0
2500	0	6	92000	0	1	4	5	0	0

CASE 8

AMERICAN HISTORY ILLUSTRATED

Imagine you are the staff assistant to the publisher of *American History Illustrated.* Imagine further that you reported yesterday to your boss, the publisher, that the renewal rate for subscriptions increased from 0.512 in January to 0.641 in February. The renewal rate is computed as the number of subscriptions renewed in a given month divided by the total number of subscriptions that expired in that month. It gives the fraction renewed of subscriptions that could have been renewed. Imagine, finally, that your boss was pleased to hear this and wanted to know which kinds of subscriptions were contributing most to the increase. He asked you to prepare a further summary by breaking the subscriptions down into several categories: gift subscriptions, previously renewed subscriptions, direct mail subscriptions, subscriptions from a subscription agency, and those from a catalog agent. This follows a standard industry classification scheme. You have obtained the data, as follows.

TABLE

	Gift subs	Previous renewal	Direct mail	Sub. agent	Catalog agent	Total subs
January						
Total	3,594	18,364	2,986	20,862	149	45,955
Renewed	2,918	14,488	1,783	4,343	13	23,545
February						
Total	884	5,140	2,224	864	45	9,157
Renewed	704	3,907	1,134	122	2	5,869

This case is based on real data, originally reported by C. H. Wagner in *The American Statistician* 36, no. 1, (February 1982), pp. 46–48. Reprinted with permission from *The American Statistician.* Copyright 1982 by the American Statistical Association. All rights reserved.

For example, the total renewal rate for January was 23,545/45,955 = .512, and the renewal rate for February for previously renewed subscriptions was 3907/5,140=0.760.

Provide the summary the boss requested, and interpret it for him.

CASE 9

HEALTHY LIFESTYLES

The Centers for Disease Control and Prevention (CDC) in Atlanta, Georgia, is the government agency responsible for disease-related issues in the United States. The CDC coordinates efforts to counteract outbreaks of diseases and funds a variety of medical and health research studies. The CDC also serves as a central clearing house for health-related data.

The CDC conducts the annual Behavioral Risk Factor Surveillance Survey (BRFSS). The survey measures a whole series of lifestyle characteristics that relate to health and longevity, such as smoking and use of seat belts. The survey compiles data on a state-by-state basis.

Your task is to prepare a summary of the latest BRFSS results. Your report is to be issued to major news organizations, such as the Associated Press, and will appear in major newspapers around the globe. For this reason, it would be inappropriate to use technical jargon in your report.

Your boss has suggested a few general ideas about what is likely to appeal to your target audience. As you study the data, you might find other things worth including.

- Report any interesting (e.g., unexpected, humorous, or odd) differences between states.
- Devise a weighted index of all seven lifestyle variables. The weighted index is to serve as an overall or composite measure of healthy lifestyles. Apply your weight to the states of Minnesota, Florida, and California as an example of what your weighted index shows.
- Discuss any noteworthy limitations of the survey or data set.

The data set in the file named HEALTHY contains six measures taken from the BRFSS. The data are described in more detail below, but feel free to visit the BRFSS website (*www.cdc.gov/nccdphp/brfss/about.htm*) for more information. All data are taken from the 1995 BRFSS except for those regarding exercise, which are taken from the 1994 BFRSS. The exercise data for Rhode Island are not available.

DEFINITIONS

BINGE — Percentage of people responding yes to the question, "Considering all types of alcoholic beverages, how many times during the past month did you have five or more drinks on an occasion?"

DWI — Percentage of people responding one or more times to the question, "During the past month, how many times have you driven when you've had perhaps too much to drink?"

BMI — Each respondent's reported height and weight were translated into a body mass index: BMI = (weight in kg)/(height in meters)^2. A BMI greater than or equal to 27.8 for men, and 27.3 for women, is considered unhealthy. The data represent the percentage of people in the unhealthy BMI range.

SMOKER — Percentage of people who have smoked at least 100 cigarettes during their lifetime and are current smokers.

SEATBELT — Percentage of people who report they do not always wear a seatbelt when they drive or ride in a car.

EXERCISE — Percentage of people responding no to the question, "During the past month, did you participate in any physical activities or exercises such as running, calisthenics, golf, gardening, or walking for exercise?"

DATA SET

STATE	BINGE	DWI	BMI	SMOKER	SEATBELT	EXERCISE
Alabama	13.58	2.62	31.83	24.53	32.26	45.81
Alaska	19.20	1.25	31.35	24.96	33.06	22.84
Arizona	13.46	2.69	24.53	22.86	25.93	23.65
Arkansas	8.75	1.49	30.14	25.20	32.65	35.10
California	15.30	1.90	26.43	15.50	14.78	21.75
Colorado	16.30	3.06	21.88	21.79	35.46	17.16
Connecticut	14.41	2.49	24.68	20.79	30.83	21.90
Delaware	8.61	1.38	29.49	25.45	29.04	36.35
Florida	13.12	2.60	29.80	23.13	24.31	27.92
Georgia	11.99	2.15	28.21	20.46	36.11	32.87
Hawaii	12.37	2.07	21.83	17.76	12.63	20.70
Idaho	12.94	2.02	27.18	19.76	42.53	21.84
Illinois	13.58	1.76	30.09	23.08	30.66	33.46
Indiana	12.75	2.58	34.65	27.19	43.33	29.62
Iowa	17.95	3.30	31.58	23.17	37.25	33.25
Kansas	13.86	3.20	27.81	22.01	42.09	34.43
Kentucky	9.70	0.59	28.79	27.83	34.75	45.86
Louisiana	14.01	2.80	30.74	25.21	32.81	33.41
Maine	11.45	0.93	26.94	24.96	48.74	40.72
Maryland	8.21	1.14	29.08	21.22	25.38	30.15

(concludes on the following page)

STATE	BINGE	DWI	BMI	SMOKER	SEATBELT	EXERCISE
Massachusetts	17.83	3.50	21.93	21.74	41.77	24.01
Michigan	18.26	3.30	31.62	25.74	29.70	23.02
Minnesota	18.01	4.89	28.35	20.48	40.82	21.74
Mississippi	8.67	1.12	31.62	24.03	41.90	38.42
Missouri	14.10	2.13	32.88	24.26	36.10	31.84
Montana	14.31	3.40	24.91	21.13	43.25	21.02
Nebraska	15.83	2.76	29.18	21.90	46.91	24.22
Nevada	18.97	3.73	26.91	26.31	28.34	21.59
New Hampshire	16.59	1.62	25.89	21.45	46.98	25.62
New Jersey	13.95	1.98	24.36	19.17	31.54	30.89
New Mexico	14.06	3.26	23.84	21.17	15.58	19.35
New York	12.39	0.90	27.83	21.47	25.80	37.12
North Carolina	5.76	1.11	28.87	25.84	14.06	42.79
North Dakota	16.97	4.19	30.67	22.69	58.31	31.98
Ohio	9.87	1.64	31.53	25.98	31.80	37.97
Oklahoma	6.69	1.17	24.14	21.66	45.93	30.38
Oregon	13.90	1.77	28.76	21.79	17.14	20.76
Pennsylvania	19.38	3.58	30.08	24.17	38.97	26.45
Rhode Island	18.70	3.65	24.90	24.74	49.87	*
South Carolina	9.17	1.39	28.65	23.74	23.20	31.36
South Dakota	14.36	5.24	28.65	21.79	56.83	30.67
Tennessee	5.22	1.01	30.95	26.46	38.20	39.66
Texas	15.27	3.70	28.60	23.66	22.35	27.80
Utah	9.93	1.22	25.04	13.16	39.66	20.90
Vermont	15.96	2.40	25.36	22.11	28.77	23.24
Virginia	14.46	2.54	29.24	22.03	27.67	22.92
Washington	13.44	2.12	25.43	20.16	22.02	18.16
West Virginia	5.92	0.87	31.94	25.73	30.01	45.28
Wisconsin	22.89	4.54	30.15	21.80	43.83	25.89
Wyoming	15.57	3.22	27.26	22.02	52.12	20.95

CASE 10

LAKE COUNTY LUNCHES

Lake County, Colorado (county seat: Leadville), is located about 100 miles west of Denver in the Colorado Rockies. The county is home to the highest airport in the continental United States; much of the county lies above 9,000 feet. The economy is largely dominated by mining (with its boom and bust cycles) and tourism.

As in many other rural areas, the school lunch program is an important component of public policy. For many poor children, the lunch they get in school provides most of their daily nutrition. Over a recent 11-year interval, the average daily number of lunches served generally declined. Since the average per capita income in the county also declined, the director of the program was eager to understand the reasons for the decline. If the families were getting poorer and simultaneously fewer lunches were being served, it would raise questions about how well the program was meeting its goals.

Eleven years' data are included in file LAKE. The variables include: YEAR, the calendar year; POP, Lake County population; UNEMPL, percentage unemployment in the state of Colorado; LKUNEMP, percentage unemployment in Lake County; LUNCH, average daily lunches served; INCOME, average per capita income in Lake County; and ENROLL, enrollment in Lake County schools. Based on these data, should the director be worried?

This case describes an actual situation. The data and the situation were brought to our attention by Carmen Lamar Arnold, Partner, Langford Partners, A. G.; Terrance L. Bartell, Assistant Business Manager, El Jebel Temple, Shriners; Jeanne Donohue; and Esther Kettering, Corporate Secretary, Sales and Acquisition Director, Golden Securities, Inc.

CASE 10 LAKE COUNTY LUNCHES

DATA SET

YEAR	POP	UNEMPL	LKUNEMP	LUNCH	INCOME	ENROLL
1	8830	5.9	9.1	940	10996	1918
2	8677	5.5	9.0	929	11728	1843
3	8523	7.7	29.2	924	10190	1630
4	7788	6.7	31.3	793	8207	1419
5	7325	5.6	16.7	746	9269	1355
6	6980	5.9	13.2	804	9814	1290
7	6667	7.4	18.2	718	9253	1217
8	6292	7.7	21.3	660	9092	1117
9	6010	6.4	15.9	725	8908	1154
10	6197	5.8	12.0	719	8515	1163
11	5975	5.0	11.0	723	8115	1147

CASE 11

MACHINE PRODUCTION RECORDS

Helen managed a major manufacturing facility that was devoted largely to producing millions of identical small metallic parts. While the parts produced were (intended to be) identical, the facility produced them by using hundreds of presses, of three different types (which we'll call types 1, 2, and 3). The rated production rates for machine types 1, 2, and 3 were 700, 200, and 155 parts per minute, respectively, but actual production rates varied. Helen felt that factors such as quality of input material, worn or "changed-out" dies, and dirty or poorly maintained presses would probably affect production.

The facility had always kept daily production records, but from what Helen could tell, no one ever consulted them or did anything with them other than to put them in file cabinets at the end of each reporting period. She thought that in principle she should be able to monitor production rates for the types of machines, identify any machines that seemed to be in need of adjustment, and characterize the amount of downtime to be expected. Accordingly, she extracted one day's production records for 116 presses at her facility.

The machines were scheduled for a shift of 7.75 hours each day. The operators recorded the hours of operation manually on clipboards kept near each machine. The actual quantities of parts produced were determined from automatic counters on the machines. The data on the next two pages (and in file PROD) were derived from these logs.

Can you advise Helen?

This case is based on a real production line situation, though the name of the organization and the exact nature of the product have been disguised. The data were provided by Ted Manzanares.

CASE 11 MACHINE PRODUCTION RECORDS

DEFINITIONS

MTYPE	Type of press (1, 2, or 3)
PROD	Production, total parts produced
HRSWRK	Hours of production time
HRSDOWN	Hours the press was down, or inoperative, for any reason

DATA SET

OBS	MTYPE	PROD	HRSWRK	HRSDOWN	OBS	MTYPE	PROD	HRSWRK	HRSDOWN
1	1	320242	7.75	0	39	1	127686	3.75	4
2	3	54311	7.75	0	40	3	60159	7.58	0.17
3	2	60820	7.75	0	41	3	48111	6	1.75
4	1	230679	5.16	2.5	42	3	22461	2.5	1.25
5	1	253855	5.67	2	43	1	308209	7.58	0.17
6	1	142151	3.57	4	44	3	82915	7.58	0.17
7	3	16118	0.58	7.17	45	2	53671	7.42	0.33
8	1	267881	7	0.75	46	2	50006	6.95	0.8
9	2	25295	4.08	3.67	47	1	214085	6	1.7
10	2	59360	7.75	0	48	2	53614	6.83	0.92
11	3	65201	6.83	0.92	49	3	82810	7.75	0
12	2	20521	3	4.75	50	1	109567	3	4.7
13	2	22916	3	4.75	51	1	245819	6.5	1.2
14	1	315124	7.75	0	52	3	74643	7.75	0
15	3	47854	6	1.75	53	3	30111	3.67	4.08
16	2	59868	7.62	0.13	54	3	30561	3.75	0
17	3	75393	6.83	0.92	55	3	47696	4.41	3.34
18	2	51374	7.25	0.5	56	2	40289	5.75	2
19	3	30372	4.34	3.41	57	1	297330	7.75	0
20	3	42784	5.75	2	58	3	31593	3.5	0.25
21	1	202003	6	1.7	59	3	46471	4.75	3
22	2	54597	7.75	0	60	3	31499	3.5	0.25
23	2	59066	7.75	0	61	3	60469	7.33	0.42
24	3	83712	7.75	0	62	3	63703	7.75	0
25	3	20530	2.5	1.25	63	2	19769	2.92	4.83
26	3	18188	2.17	5.58	64	1	212743	7.75	0
27	3	87400	7.75	0	65	2	55767	7.75	0
28	1	311877	7.75	0	66	3	15038	1.75	2
29	2	15783	1.75	6	67	3	31556	3.5	0.25
30	1	259945	7.58	0.17	68	3	30380	3.42	0.33
31	2	31034	4.83	2.92	69	3	68982	6.58	1.17
32	3	25569	4.17	3.58	70	2	59850	7.7	0.05
33	2	15771	2.67	5.08	71	1	300891	7.75	0
34	2	6899	1.5	6.25	72	2	56431	7.75	0
35	3	82837	7.75	0	73	3	83171	7.75	0
36	2	58349	7.75	0	74	3	16650	2.25	5.5
37	3	68899	7.75	0	75	2	52644	7.43	0.32
38	3	41909	5.25	2.5	76	3	8954	1.08	2.67

(concludes on the following page)

OBS	MTYPE	PROD	HRSWRK	HRSDOWN	OBS	MTYPE	PROD	HRSWRK	HRSDOWN
77	1	316080	7.75	0	97	3	61092	7.5	0.25
78	2	20200	2.93	4.82	98	3	87114	7.75	0
79	2	61050	7.65	0.1	99	3	8675	1.08	2.67
80	2	30459	4.47	3.28	100	1	301077	7.5	0.2
81	2	25061	2.83	4.92	101	3	5084	0.58	3.17
82	2	60224	7.75	0	102	2	55753	7.62	0.13
83	3	60935	7.5	0.58	103	1	285831	7.75	0
84	3	31233	3.75	0	104	1	302774	7.75	0
85	3	20237	1.92	5.83	105	2	57083	7.25	0.5
86	1	283785	7.25	0.5	106	1	281949	7.25	0.5
87	2	49499	6.75	1	107	3	40469	6.08	1.67
88	3	59722	6.67	1.08	108	1	300039	7.75	0
89	1	271276	6.58	1.1	109	3	88634	7.67	0.08
90	3	62762	7.75	0	110	1	312501	7.75	0
91	2	61075	7.75	0	111	3	75455	6.83	0.92
92	1	220686	6.83	0.92	112	2	27600	3.63	4.12
93	3	78103	7.75	0	113	1	273486	7	0.75
94	2	51300	7.54	0.21	114	3	52164	6.67	1.08
95	3	40967	3.25	4.5	115	2	29367	5.05	2.7
96	3	31056	3.5	0.25	116	3	8142	1.08	2.67

CASE **12**

KERRVILLE RAINFALL

Everybody talks about the weather, but no one *does* anything about it, according to Mark Twain. The data on the next page give the total annual rainfall (TOTAL, in inches) for Kerrville, Texas, from 1931 to 1995. Summarize any patterns you find in this time series. Can these data be used to predict rainfall for Kerrville for 1996? These data are also available in file KERRVILL.

The data used in the case are from the official NOAA-NWS supervised Cooperative Weather Observation Site for Kerrville, Texas, located at the Knipling-Bushland U.S. Livestock Insects Research Laboratory. They were collected by volunteer weather observers, including Al Siebenaler, in cooperation with NOAA.

CASE 12 KERRVILLE RAINFALL

DATA SET

YEAR	TOTAL	YEAR	TOTAL
1931	35.07	1964	31.85
1932	41.53	1965	40.86
1933	19.14	1966	27.62
1934	24.22	1967	27.84
1935	49.29	1968	33.33
1936	47.68	1969	32.03
1937	27.10	1970	21.43
1938	20.95	1971	34.48
1939	28.42	1972	28.39
1940	39.14	1973	33.33
1941	39.66	1974	38.87
1942	29.16	1975	29.29
1943	21.38	1976	34.71
1944	36.87	1977	23.57
1945	33.89	1978	44.26
1946	34.48	1979	42.31
1947	27.19	1980	27.45
1948	25.55	1981	41.51
1949	41.82	1982	21.54
1950	22.70	1983	25.14
1951	18.15	1984	22.48
1952	40.91	1985	36.03
1953	26.36	1986	38.19
1954	14.64	1987	42.13
1955	28.90	1988	30.90
1956	14.04	1989	23.16
1957	55.09	1990	33.58
1958	36.71	1991	44.52
1959	29.30	1992	41.40
1960	37.07	1993	23.06
1961	26.63	1994	38.76
1962	17.13	1995	28.26
1963	21.34		

CASE 13

WORKLOAD RATINGS

Many organizations use opinion surveys as one part of their evaluations of managers or executives. The managers being evaluated often wonder how fair these processes are. In particular, they wonder if they will somehow get lower ratings if they strictly enforce rules or set high standards for work.

In universities the same problem affects instructors, who complain about the student evaluations of the instructors and their courses. In particular, they complain that when instructors make students work hard, the students give them low ratings. Is this really true? (The students, in turn, often wonder if anyone ever looks at the evaluations they fill out, but that's another story.)

All of the classes offered in a recent fall semester by the University of Colorado at Denver, College of Business and Administration, were evaluated by the students in those classes, by using a standard form prescribed by the University's Board of Regents. Students responded to 12 questions, each of which was scored from 0 to 4 (0 = "F," 4 = "A"), and the class averages for the responses to each question were tabulated, provided to the instructor, and given to deans, department chairs, and to the student government, which published them for general review.

For each of the 161 sections, the class averages for three of the questions are listed on the next two pages (and are available in file WORKLOAD). The variables on that file are as follows. CRSRTG is the course rating: "Rate this course compared with your other university courses." WORKLOAD is the workload: "Rate the work required for this course" (F = too little, C = about right, A = too much). Finally, INSTRTG is the instructor rating: "Rate this instructor compared to your other university instructors."

Note that for WORKLOAD, a rating of 2.0 would be "about right," as perceived by the students. For CRSRTG and INSTRTG, 0 means a failing rating, 4 means a perfect one. All of the numbers given are class averages for the particular sections offered.

Do WORKLOAD and INSTRTG appear to be related, based on these data? Do they appear related if we adjust first for the effects of the course rating? Do these data support the rumors about instructors' being penalized for making students work hard?

CASE 13 WORKLOAD RATINGS

WORKLOAD	CRSRTG	INSTRTG	WORKLOAD	CRSRTG	INSTRTG
2.33	2.48	2.19	3.24	2.43	2.72
2.33	3.17	3.42	3.00	2.41	2.56
3.07	3.60	3.53	3.17	2.78	3.00
2.19	3.15	3.24	2.58	3.58	3.50
2.04	3.53	3.68	1.50	1.69	1.94
2.34	3.29	3.36	2.29	3.23	3.40
2.11	2.28	2.11	2.46	3.32	3.68
2.11	2.11	1.78	2.03	2.97	3.57
1.67	2.00	1.33	2.14	3.13	3.46
2.33	3.83	3.75	2.12	2.87	2.87
1.95	2.91	3.23	2.08	3.23	3.62
2.19	3.00	3.12	3.00	3.55	3.75
1.73	2.62	2.75	2.00	3.13	3.47
2.95	3.05	2.84	2.43	2.76	3.35
2.78	2.51	2.62	2.05	2.55	2.72
2.65	2.69	2.62	2.54	3.69	3.69
2.50	3.25	3.33	2.00	2.80	3.15
2.29	3.75	3.75	2.08	3.08	3.37
1.88	3.03	3.18	2.37	3.62	3.69
1.93	2.79	3.00	3.11	3.67	.67
2.15	3.62	3.87	2.39	3.00	3.24
2.04	3.22	3.52	2.14	2.81	2.86
2.78	2.78	3.35	2.17	3.17	3.34
2.73	2.77	3.00	2.40	3.40	3.56
2.29	2.71	2.43	2.04	2.81	2.96
2.93	3.27	3.19	2.22	3.03	3.12
3.56	2.29	1.96	1.95	2.70	3.05
2.65	3.50	3.57	2.25	2.50	2.25
3.11	3.52	3.59	2.73	3.50	3.82
1.83	2.00	2.00	2.34	1.90	2.03
2.90	3.75	3.65	2.50	2.00	2.24
2.12	3.62	3.69	2.32	2.31	2.31
2.42	3.23	3.19	2.04	2.04	2.00
2.21	3.43	3.43	2.09	3.48	3.57
2.12	3.50	3.62	2.21	1.89	1.63
2.17	2.68	2.88	2.09	1.81	1.74
2.10	2.61	2.77	2.72	3.53	3.57
2.00	3.83	3.67	3.00	3.13	3.13
2.40	2.75	2.75	2.50	2.92	2.83
1.96	2.48	2.83	2.75	3.33	3.62
1.87	2.50	2.63	3.39	2.86	3.07
1.94	2.50	2.57	2.22	3.70	3.74
2.54	2.73	2.54	2.10	3.50	3.50
2.84	2.94	3.19	2.02	2.95	2.98

(concludes on the next page)

CASE 13 WORKLOAD RATINGS

WORKLOAD	CRSRTG	INSTRTG	WORKLOAD	CRSRTG	INSTRTG
2.36	2.51	2.53	1.94	3.06	3.21
2.62	2.04	2.19	2.15	2.63	2.81
2.09	2.82	2.77	2.12	2.36	3.07
2.07	2.89	2.93	2.16	3.48	3.68
2.17	3.08	3.00	2.00	3.55	2.72
2.85	2.45	2.63	2.09	2.90	3.03
1.96	3.83	3.92	2.13	1.67	1.33
2.11	3.21	3.29	1.95	2.71	2.43
2.06	2.59	2.88	1.95	2.00	2.10
2.08	3.42	3.62	2.87	2.27	2.59
2.37	2.81	3.00	2.16	3.20	3.40
2.30	2.77	3.18	2.20	3.52	3.73
2.37	2.52	3.00	2.17	2.90	3.13
2.43	2.64	2.91	1.82	2.80	3.09
2.02	1.75	1.80	2.22	3.75	3.62
2.35	2.55	2.45	1.86	2.64	2.41
2.52	2.84	2.90	3.65	2.12	2.19
3.74	2.67	2.59	2.00	3.50	4.00
2.18	2.18	1.96	2.31	3.54	3.46
2.03	2.40	3.03	2.72	.89	1.11
2.09	2.64	3.27	2.26	2.57	2.43
2.17	1.67	2.00	2.37	1.80	1.83
2.30	3.40	3.70	2.38	2.62	3.23
2.22	2.82	3.35	3.47	2.07	2.40
2.68	2.87	3.09	3.65	2.45	2.40
3.04	2.78	3.00	3.00	2.89	3.00
3.29	3.00	2.88	2.36	1.82	1.83
3.25	2.68	3.07	2.10	3.57	3.52
2.00	2.67	2.67	2.07	3.38	3.37
1.73	2.91	2.91	2.12	2.39	2.42
2.14	2.33	2.14	2.43	3.57	3.71
2.26	2.77	3.17	2.29	3.71	3.71
2.18	1.59	1.76	1.97	3.44	3.50
2.32	3.26	3.45	2.17	1.52	1.48
2.22	3.41	3.68	2.50	2.40	2.07
2.17	3.31	3.47	2.78	2.00	1.91
1.33	1.44	1.72			

CASE **14**

ROARING FORK TRANSIT

The Roaring Fork Transit Agency (RFTA) operates a multicounty bus system based in Aspen, Colorado, a mountain community known as an upscale resort. The Roaring Fork Valley (named for the Roaring Fork River, which flows from the mountains above Aspen into the Colorado River at Glenwood Springs, some 50 miles away) grew substantially from 1987 to 1990. Combined with lack of available affordable housing (particularly in the winter ski season), increased traffic on state highway 82 ("Killer 82," the only highway into Aspen in winter), and rising gasoline prices ($1.83 per gallon in January 1991), the growth led to more passengers on the "downvalley" routes from Aspen to El Jebel and Carbondale. At times, RFTA had trouble meeting demand for its services on this route.

Monthly ridership figures (total riders) for OUTBOUND (from Aspen in the direction of Carbondale/El Jebel) and INBOUND (from Carbondale/El Jebel to Aspen) routes are given in the data below for 1987–1990 (months are indicated by the variable MONTH, where MONTH=1 for January 1987).

In early 1991, RFTA management asked an analyst to advise them on choices about buying new equipment, leasing extra buses, hiring more drivers, and so forth, for the next few years. Prepare a presentation highlighting any trends or patterns in the data you think the analyst and RFTA management should consider. The data are available in the file ROARING.

This case is based on a real situation. It was described to us and the data were provided by Rebecca R. Horton and Cheryl Cumnock, Personnel Director, Pitkin County, Colorado.

CASE 14 ROARING FORK TRANSIT

DATA SET

MONTH	OUTBOUND	INBOUND
1	8078	7885
2	7816	7004
3	8025	7058
4	5170	4635
5	3534	3130
6	4449	4242
7	6508	6046
8	5849	5250
9	4262	3759
10	3913	3377
11	4666	4174
12	8534	7995
13	11028	12114
14	11517	11190
15	12012	12109
16	6577	6030
17	4677	4477
18	7583	7144
19	10947	10720
20	11590	10823
21	5652	5086
22	4829	4164
23	7918	7608
24	23536	19697
25	12991	13353
26	11999	12752
27	11986	12151
28	6206	7472
29	6787	6481
30	9432	9463
31	12826	13240
32	12626	12826
33	8223	8059
34	7647	7414
35	10652	10941
36	18767	19889
37	19815	19747
38	17078	17124
39	17662	18030
40	10429	10670
41	8595	8672
42	11478	11968
43	16635	16394
44	16638	16336
45	12500	12644
46	12725	13088
47	15745	17011
48	26362	29890

CASE **15**

STURGEL DIVISION

"Now that I write it all down, I see we have changed a lot!" mused Martha as she put the finishing touches on her annual status report. "In fact," she went on, "the name *information services* indicates most of the changes. We used to be *information systems.*"

Martha was information services (IS) manager for the Sturgel Division of a major manufacturing company. Sturgel developed and manufactured (mostly small) household appliances on a 230-acre site in the southeast part of the United States. While she had managed the IS department for only 15 months, she had been part of the department for nine years, all at the same location.

The late 1990s were a time of many changes for the information systems departments in most companies, including the Sturgel Division. Technological change drove most of the organizational change: the price performance of most computer equipment had improved by compound rates of perhaps 30% per year for several decades. This meant that many companies no longer needed a "computer center." Individual users could simply do whatever they needed to do on a desktop computer in their own offices.

Martha made a rough list of the major influences she'd dealt with over the last few years, calling them "IS Transitions." They included:

• From running the computer to providing information services to the company.
• From "owning" the data to consulting with user departments about use of the company's information resources.
• From emphasizing mainframes to emphasizing terminals, personal computers, and telecommunications. Indeed, sometimes it seemed that the telephone, word processing, clerical, and library departments had all been combined.
• From developing the company's applications to advising users on how to develop their own applications.

Each transition had been difficult, for each required its own combination of hardware, software, data, procedures, and people, most of which were different at the end of the transition.

Currently, the Information Service Department had four sections, each with a manager who reported to Martha:

1. Systems operations. The systems operations section ran two mainframe computers and one minicomputer that operated as a telecommunications "node" and link to other divisions and corporate headquarters. They provided operator support for three shifts of operation, as well as systems maintenance, operating system updates, and so forth. Hardware maintenance was handled by the vendor of the mainframes.

2. Application development. The application development section really should be renamed, Martha thought. Most of its activities involved data base design and maintenance, though two small groups developed financially oriented applications and maintained several software packages aimed at serving the (fairly small) engineering staff at Sturgel.

3. PC services. The PC services section developed division standards for personal computers at Sturgel, maintained the hardware, and consulted with users who wanted PCs of their own. The users purchased PCs out of their own budgets, but they were required to meet certain hardware and software standards as set by IS. This section also evaluated general-purpose software, such as word processing and spreadsheet packages, and recommended to users when they should switch to a new version or release. They also did a fair amount of "hand holding" for users who had difficulty developing their own applications. Recently, the section had developed new procedures for nationwide maintenance (and security) of the company's laptop computers.

4. Telecommunications. The telecommunications section maintained the local area network at the Sturgel site, as well as links to corporate headquarters and other company divisions. They also advised users on such activities as links to legal databases and information retrieval services, though that activity, thought Martha, might really fit better in the application development section.

Martha felt good about the organization in general. While it had been a bit of a scramble to develop an organization that could deal with the fast pace of technological change in the computer field, she felt they had more or less done it. One part of it still made her nervous, though. The biggest theme in all the changes, from an IS perspective, was the change from computing and running computers to serving large numbers of people directly.

She felt they were providing good service, but she had no regular way of knowing how her users perceived it. She thought she should consider a regular survey of her users and their perceptions of service received from the IS department. That way, she figured, she would be in a position to have data to back up her informal sense, and she would (presumably) learn about changes in those perceptions (for better or for worse) more quickly.

What kind of survey should Martha run, or should she? How should it be administered? What kind of forms, questionnaires, or other survey approaches should be developed? Martha would like you to develop an appropriate scheme.

CASE 16

PHOTOCOPYING ABUSES

Your firm is located in a 10-story building. Each floor has its own copy machine. The firm owns these machines but must pay for paper, toner, and maintenance. Each employee has a key that opens the copy room door on his/her floor only and does not have access to copy machines on other floors.

Because the copy machines are "free goods" right now, you suspect that the firm's photocopying costs could be cut pretty drastically. To test this theory, you have decided to perform an experiment. Each person on the tenth floor has been given a card that operates the tenth-floor machine. These employees have been told that their cards will generate a daily accounting of their photocopying activity. Tenth-floor employees have also been told that they will not be *charged* for their use of the machine, but they certainly know that *someone* will have some sense of individual usage patterns.

To establish a basis of comparison, the group on the third floor has not been converted to the card system. The third-floor machine has an internal mechanism that totals the number of copies made each day, but you do not know *who* is doing *what,* and the third-floor employees have no reason to believe that they are being monitored. The third and tenth floors have been chosen for this experiment because these two floors have had about the same usage rates in the past.

You have collected data from the two machines over the last 50 working days. The data are in the file named COPIERS. This file has 50 observations and three variables: DAY, an indicator of the day; TENTH, the number of copies made on the tenth floor; and THIRD, the number of copies made on the third floor.

Use appropriate graphical or statistical techniques to determine whether the card accounting system will effectively lower photocopying usage if implemented companywide.

The scenario described in this case was related to us by one of our students, who was employed at the firm when the copy machine experiment was undertaken.

CASE 16 PHOTOCOPYING ABUSES

DATA SET

DAY	TENTH	THIRD	DAY	TENTH	THIRD
1	500	440	26	250	320
2	420	220	27	250	360
3	440	360	28	230	450
4	480	110	29	240	270
5	450	240	30	220	380
6	460	360	31	190	190
7	450	80	32	150	500
8	420	420	33	170	290
9	410	310	34	120	150
10	405	30	35	150	370
11	380	290	36	140	405
12	360	410	37	130	130
13	360	460	38	150	120
14	370	420	39	130	70
15	350	150	40	110	240
16	320	170	41	90	20
17	350	250	42	80	450
18	360	20	43	90	20
19	310	250	44	70	40
20	320	350	45	20	320
21	290	150	46	50	140
22	290	250	47	40	90
23	270	230	48	20	130
24	250	90	49	30	480
25	240	50	50	30	350

CASE 17

AGRICOMP

"How do I make sense of all this?" asked Jody, as he prepared to write his report to the dealer relations committee. He stared at the 292 responses to a survey of AgriComp's dealers as he began to ponder the problem. The question was whether to recommend a change to AgriComp's current procedures for settling warranty claim disputes with its dealers.

AgriComp sold computer systems to farmers, who used the systems for such purposes as crop rotation planning and spreadsheet analysis for financial planning. Many also used the systems to provide remote access to agriculture-oriented data bases, market news, and even weather information. The equipment was assembled at company headquarters in southern Minnesota. The software was provided by subcontractors but was distributed under AgriComp's name. Both the hardware and software were sold through some 350 affiliated dealers nationwide, 292 of whom had responded to Jody's survey. It was relations with these dealers that concerned him.

The local dealers handled warranty service for AgriComp products. When hardware or software problems occurred, they arranged for appropriate repairs to be made locally and submitted vouchers to AgriComp headquarters in Minnesota. The headquarters staff reviewed the vouchers and issued reimbursement checks to the dealer. Occasionally claims were denied when the staff found that the particular repair was not covered by the company's warranty or that the warranty had expired. In such cases, the dealer was more or less stuck for the cost of the repairs, and this had caused occasional hard feelings.

The company had an internal appeals process for dealers to follow to protest such denials but, at the last dealers' meeting, Jody had heard a lot of grumbling about that

This case was inspired by an actual survey taken during 1993, but we have disguised the nature of the organization and its products, as well as the particular issue in this survey.

process. More than one dealer had suggested that it was useless, as appeals were always denied. The costs of repairs might correspond to the profits on many systems for the dealer, so their concern was understandable.

Jody knew that clearer warranty instructions would help. Sometimes the dealers couldn't understand exactly what was or wasn't covered by warranty. In that kind of case, they often arranged for the repairs (to keep the customer happy) and took their chances on reimbursement. New documentation currently being developed would probably help with that part of the problem.

In a corridor conversation at the dealers' meeting, one dealer had suggested that perhaps in cases of dispute, an impartial mediator, external to the company, might be called in to settle the matter. In the annual survey of dealers, Jody had included a question about that proposal. As part of a one-page survey, dealers were asked to respond to the statement "The warranty appeals process should be replaced by a process using impartial mediators" on a scale of 1 to 5, where 1 indicated "Strongly agree," 3 indicated "Neither agree nor disagree," and 5 indicated "Strongly disagree." Each dealer was also asked to give the number of times in the last three years in which he or she had used the warranty appeals process. Answers to this question were 0, 1, 2, and "3 or more."

Part of the data from that survey is shown below, and is also available in file AGRICOMP. REP is the dealer's response to the question about a new appeals system and USE is the number of times the dealer used the appeals process, where 3 indicates "3 or more."

Jody was willing to consider changing the warranty appeals process along the lines suggested, if it were important to the dealers. It would cost the company some money, both for the external mediator and (perhaps) for increased costs from appeals the company lost, but keeping the dealers happy was obviously important. Jody wondered, though, just how important it was to the dealers? Was there widespread discontent, or had he just heard from a few malcontents at the dealers' meeting?

DATA SET

REP	USE	REP	USE	REP	USE	REP	USE	REP	USE
5	3	3	0	3	2	4	3	2	1
5	2	5	3	3	1	5	3	4	3
5	1	2	2	2	1	4	2	3	2
5	2	4	3	4	2	5	3	2	2
2	2	5	1	1	1	5	2	3	2
4	3	4	3	3	0	4	3	4	1
1	0	4	3	5	1	2	2	4	3
3	3	4	0	4	3	2	0	4	3
2	2	5	3	5	1	1	0	1	2
4	2	5	1	4	0	2	3	5	2
4	3	3	1	1	0	2	2	4	3
1	0	2	1	1	1	5	3	5	1
2	1	3	1	2	1	4	1	4	3
3	1	5	3	2	1	4	3	4	3
4	2	2	2	1	2	2	2	4	3
3	0	2	3	3	3	3	1	4	1
3	1	2	2	4	3	2	1	2	1

(concludes on the following page)

CASE 17 AGRICOMP I-49

REP	USE	REP	USE	REP	USE	REP	USE	REP	USE
3	2	3	2	4	3	3	2	2	1
3	3	2	1	1	0	4	3	2	3
2	1	3	3	2	1	2	2	3	3
5	3	2	3	2	2	3	3	2	1
5	3	4	2	3	2	1	0	1	1
4	2	2	1	4	3	1	1	4	0
1	3	5	3	4	3	5	2	1	3
4	3	1	1	4	3	3	3	3	2
5	1	3	2	4	1	4	3	5	3
1	1	5	3	4	1	4	3	5	3
1	0	5	2	3	1	3	1	3	3
4	0	2	1	2	2	4	1	4	3
4	3	4	2	5	3	4	3	3	1
4	3	5	1	2	3	2	0	3	0
3	0	2	3	4	1	4	3	1	1
1	3	5	3	5	3	3	3	5	2
2	3	2	1	2	2	4	3	4	3
4	1	5	3	4	3	5	3	2	3
4	3	2	1	4	3	3	2	4	3
2	2	5	3	2	0	3	2	5	1
3	3	3	2	5	1	5	3	3	0
4	3	3	3	2	2	4	3	1	3
5	3	4	3	1	1	5	2	1	3
3	3	5	3	5	2	5	2	2	2
3	1	5	3	3	3	5	1	5	3
2	1	2	1	3	1	1	1	2	3
2	2	4	3	5	3	4	3	4	3
5	3	3	3	2	2	2	3	4	3
4	3	5	3	1	2	3	2	3	1
4	3	1	1	5	3	1	0	4	3
4	3	5	2	2	1	3	3	5	2
2	0	5	1	4	2	3	1	1	3
4	3	1	0	2	1	5	3	2	3
3	1	1	0	3	0	2	3	1	0
5	3	4	3	3	1	2	1	2	1
4	3	5	3	4	1	5	3	5	3
5	0	5	3	3	3	3	3	3	1
2	1	5	3	3	2	3	0	3	2
1	2	4	3	3	3	2	1	3	1
3	2	3	1	4	3	2	0		
5	3	2	0	1	0	1	1		
2	1	5	3	2	1	1	1		

CASE 18

TIPS

Foodservers' tips in restaurants may be influenced by many factors, including the nature of the restaurant, size of the party, table location in the restaurant, and so forth. To make appropriate assignments for the foodservers, restaurant managers need to know what these factors are. They must avoid either the substance or appearance of unfair treatment of the foodservers, for whom the tips are a major component of pay. In one restaurant, a foodserver recorded the following data on all customers he had served during a recent interval of two and a half months. The restaurant, located in a suburban shopping mall, was one of a national chain and served a varied menu. Pursuant to local law, the restaurant offered seating in a nonsmoking section to patrons who requested it. The data were recorded on those days and during those times when the foodserver was routinely assigned to work.

The restaurant manager thought the data might provide a means of assessing what sorts of things affected tips and in a way which might be seen as objective. The server had worked at a variety of assignments on different days and times during the data-gathering period and had been happy to cooperate. Are there any patterns of which you think the foodserver or the restaurant manager should be aware, either in the total bills or in the tips? What other variables might be useful, if the experiment were to be repeated?

The data are available in file TIPS.

This case describes a real situation. The data were provided by Thomas J. Kientz, President, Colorado National Bank, Aurora, Colorado, and by Sean Schneider.

CASE 18 TIPS

DEFINITIONS

TOTBILL	Total bill, including tax, in dollars
TIP	Tip in dollars
SEX	Sex of person paying bill (0 = male, 1 = female)
SMOKER	Smoker in party? (0 = no, 1 = yes)
DAY	3 = Thursday 5 = Saturday
	4 = Friday 6 = Sunday
TIME	(0 = day, 1 = night)
SIZE	Size of party

DATA SET

OBS	TOTBILL	TIP	SEX	SMOKER	DAY	TIME	SIZE	OBS	TOTBILL	TIP	SEX	SMOKER	DAY	TIME	SIZE
1	16.99	1.01	1	0	6	1	2	35	17.78	3.27	0	0	5	1	2
2	10.34	1.66	0	0	6	1	3	36	24.06	3.60	0	0	5	1	3
3	21.01	3.50	0	0	6	1	3	37	16.31	2.00	0	0	5	1	3
4	23.68	3.31	0	0	6	1	2	38	16.93	3.07	1	0	5	1	3
5	24.59	3.61	1	0	6	1	4	39	18.69	2.31	0	0	5	1	3
6	25.29	4.71	0	0	6	1	4	40	31.27	5.00	0	0	5	1	3
7	8.77	2.00	0	0	6	1	2	41	16.04	2.24	0	0	5	1	3
8	26.88	3.12	0	0	6	1	4	42	17.46	2.54	0	0	6	1	2
9	15.04	1.96	0	0	6	1	2	43	13.94	3.06	0	0	6	1	2
10	14.78	3.23	0	0	6	1	2	44	9.68	1.32	0	0	6	1	2
11	10.27	1.71	0	0	6	1	2	45	30.40	5.60	0	0	6	1	4
12	35.26	5.00	1	0	6	1	4	46	18.29	3.00	0	0	6	1	2
13	15.42	1.57	0	0	6	1	2	47	22.23	5.00	0	0	6	1	2
14	18.43	3.00	0	0	6	1	4	48	32.40	6.00	0	0	6	1	4
15	14.83	3.02	1	0	6	1	2	49	28.55	2.05	0	0	6	1	3
16	21.58	3.92	0	0	6	1	2	50	18.04	3.00	0	0	6	1	2
17	10.33	1.67	1	0	6	1	3	51	12.54	2.50	0	0	6	1	2
18	16.29	3.71	0	0	6	1	3	52	10.29	2.60	1	0	6	1	2
19	16.97	3.50	1	0	6	1	3	53	34.81	5.20	1	0	6	1	4
20	20.65	3.35	0	0	5	1	3	54	9.94	1.56	0	0	6	1	2
21	17.92	4.08	0	0	5	1	2	55	25.56	4.34	0	0	6	1	4
22	20.29	2.75	1	0	5	1	2	56	19.49	3.51	0	0	6	1	2
23	15.77	2.23	1	0	5	1	2	57	38.01	3.00	0	1	5	1	4
24	39.42	7.58	0	0	5	1	4	58	26.41	1.50	1	0	5	1	2
25	19.82	3.18	0	0	5	1	2	59	11.24	1.76	0	1	5	1	2
26	17.81	2.34	0	0	5	1	4	60	48.27	6.73	0	0	5	1	4
27	13.37	2.00	0	0	5	1	2	61	20.29	3.21	0	1	5	1	2
28	12.69	2.00	0	0	5	1	2	62	13.81	2.00	0	1	5	1	2
29	21.70	4.30	0	0	5	1	2	63	11.02	1.98	0	1	5	1	2
30	19.65	3.00	1	0	5	1	2	64	18.29	3.76	0	1	5	1	4
31	9.55	1.45	0	0	5	1	2	65	17.59	2.64	0	0	5	1	3
32	18.35	2.50	0	0	5	1	4	66	20.08	3.15	0	0	5	1	3
33	15.06	3.00	1	0	5	1	2	67	16.45	2.47	1	0	5	1	2
34	20.69	2.45	1	0	5	1	4	68	3.07	1.00	1	1	5	1	1

(continues on the following page)

OBS	TOTBILL	TIP	SEX	SMOKER	DAY	TIME	SIZE
69	20.23	2.01	0	0	5	1	2
70	15.01	2.09	0	1	5	1	2
71	12.02	1.97	0	0	5	1	2
72	17.07	3.00	1	0	5	1	3
73	26.86	3.14	1	1	5	1	2
74	25.28	5.00	1	1	5	1	2
75	14.73	2.20	1	0	5	1	2
76	10.51	1.25	0	0	5	1	2
77	17.92	3.08	0	1	5	1	2
78	27.20	4.00	0	0	3	0	4
79	22.76	3.00	0	0	3	0	2
80	17.29	2.71	0	0	3	0	2
81	19.44	3.00	0	1	3	0	2
82	16.66	3.40	0	0	3	0	2
83	10.07	1.83	1	0	3	0	1
84	32.68	5.00	0	1	3	0	2
85	15.98	2.03	0	0	3	0	2
86	34.83	5.17	1	0	3	0	4
87	13.03	2.00	0	0	3	0	2
88	18.28	4.00	0	0	3	0	2
89	24.71	5.85	0	0	3	0	2
90	21.16	3.00	0	0	3	0	2
91	28.97	3.00	0	1	4	1	2
92	22.49	3.50	0	0	4	1	2
93	5.75	1.00	1	1	4	1	2
94	16.32	4.30	1	1	4	1	2
95	22.75	3.25	1	0	4	1	2
96	40.17	4.73	0	1	4	1	4
97	27.28	4.00	0	1	4	1	2
98	12.03	1.50	0	1	4	1	2
99	21.01	3.00	0	1	4	1	2
100	12.46	1.50	0	0	4	1	2
101	11.35	2.50	1	1	4	1	2
102	15.38	3.00	1	1	4	1	2
103	44.30	2.50	1	1	5	1	3
104	22.42	3.48	1	1	5	1	2
105	20.92	4.08	1	0	5	1	2
106	15.36	1.64	0	1	5	1	2
107	20.49	4.06	0	1	5	1	2
108	25.21	4.29	0	1	5	1	2
109	18.24	3.76	0	0	5	1	2
110	14.31	4.00	1	1	5	1	2
111	14.00	3.00	0	0	5	1	2
112	7.25	1.00	1	0	5	1	1
113	38.07	4.00	0	0	6	1	3
114	23.95	2.55	0	0	6	1	2
115	25.71	4.00	1	0	6	1	3
116	17.31	3.50	1	0	6	1	2
117	29.93	5.07	0	0	6	1	4
118	10.65	1.50	1	0	3	0	2
119	12.43	1.80	1	0	3	0	2
120	24.08	2.92	1	0	3	0	4
121	11.69	2.31	0	0	3	0	2
122	13.42	1.68	1	0	3	0	2
123	14.26	2.50	0	0	3	0	2
124	15.95	2.00	0	0	3	0	2
125	12.48	2.52	1	0	3	0	2
126	29.80	4.20	1	0	3	0	6
127	8.52	1.48	0	0	3	0	2
128	14.52	2.00	1	0	3	0	2
129	11.38	2.00	1	0	3	0	2
130	22.82	2.18	0	0	3	0	3
131	19.08	1.50	0	0	3	0	2
132	20.27	2.83	1	0	3	0	2
133	11.17	1.50	1	0	3	0	2
134	12.26	2.00	1	0	3	0	2
135	18.26	3.25	1	0	3	0	2
136	8.51	1.25	1	0	3	0	2
137	10.33	2.00	1	0	3	0	2
138	14.15	2.00	1	0	3	0	2
139	16.00	2.00	0	1	3	0	2
140	13.16	2.75	1	0	3	0	2
141	17.47	3.50	1	0	3	0	2
142	34.30	6.70	0	0	3	0	6
143	41.19	5.00	0	0	3	0	5
144	27.05	5.00	1	0	3	0	6
145	16.43	2.30	1	0	3	0	2
146	8.35	1.50	1	0	3	0	2
147	18.64	1.36	1	0	3	0	3
148	11.87	1.63	1	0	3	0	2
149	9.78	1.73	0	0	3	0	2
150	7.51	2.00	0	0	3	0	2
151	14.07	2.50	0	0	6	1	2
152	13.13	2.00	0	0	6	1	2
153	17.26	2.74	0	0	6	1	3
154	24.55	2.00	0	0	6	1	4
155	19.77	2.00	0	0	6	1	4
156	29.85	5.14	1	0	6	1	5
157	48.17	5.00	0	0	6	1	6
158	25.00	3.75	1	0	6	1	4
159	13.39	2.61	1	0	6	1	2
160	16.49	2.00	0	0	6	1	4
161	21.50	3.50	0	0	6	1	4
162	12.66	2.50	0	0	6	1	2

(concludes on the following page)

OBS	TOTBILL	TIP	SEX	SMOKER	DAY	TIME	SIZE
163	16.21	2.00	1	0	6	1	3
164	13.81	2.00	0	0	6	1	2
165	17.51	3.00	1	1	6	1	2
166	24.52	3.48	0	0	6	1	3
167	20.76	2.24	0	0	6	1	2
168	31.71	4.50	0	0	6	1	4
169	10.59	1.61	1	1	5	1	2
170	10.63	2.00	1	1	5	1	2
171	50.81	10.00	0	1	5	1	3
172	15.81	3.16	0	1	5	1	2
173	7.25	5.15	0	1	6	1	2
174	31.85	3.18	0	1	6	1	2
175	16.82	4.00	0	1	6	1	2
176	32.90	3.11	0	1	6	1	2
177	17.89	2.00	0	1	6	1	2
178	14.48	2.00	0	1	6	1	2
179	9.60	4.00	1	1	6	1	2
180	34.63	3.55	0	1	6	1	2
181	34.65	3.68	0	1	6	1	4
182	23.33	5.65	0	1	6	1	2
183	45.35	3.50	0	1	6	1	3
184	23.17	6.50	0	1	6	1	4
185	40.55	3.00	0	1	6	1	2
186	20.69	5.00	0	0	6	1	5
187	20.90	3.50	1	1	6	1	3
188	30.46	2.00	0	1	6	1	5
189	18.15	3.50	1	1	6	1	3
190	23.10	4.00	0	1	6	1	3
191	15.69	1.50	0	1	6	1	2
192	19.81	4.19	1	1	3	0	2
193	28.44	2.56	0	1	3	0	2
194	15.48	2.02	0	1	3	0	2
195	16.58	4.00	0	1	3	0	2
196	7.56	1.44	0	0	3	0	2
197	10.34	2.00	0	1	3	0	2
198	43.11	5.00	1	1	3	0	4
199	13.00	2.00	1	1	3	0	2
200	13.51	2.00	0	1	3	0	2
201	18.71	4.00	0	1	3	0	3
202	12.74	2.01	1	1	3	0	2
203	13.00	2.00	1	1	3	0	2
204	16.40	2.50	1	1	3	0	2
205	20.53	4.00	0	1	3	0	4
206	16.47	3.23	1	1	3	0	3
207	26.59	3.41	0	1	5	1	3
208	38.73	3.00	0	1	5	1	4
209	24.27	2.03	0	1	5	1	2
210	12.76	2.23	1	1	5	1	2
211	30.06	2.00	0	1	5	1	3
212	25.89	5.16	0	1	5	1	4
213	48.33	9.00	0	0	5	1	4
214	13.27	2.50	1	1	5	1	2
215	28.17	6.50	1	1	5	1	3
216	12.90	1.10	1	1	5	1	2
217	28.15	3.00	0	1	5	1	5
218	11.59	1.50	0	1	5	1	2
219	7.74	1.44	0	1	5	1	2
220	30.14	3.09	1	1	5	1	4
221	12.16	2.20	0	1	4	0	2
222	13.42	3.48	1	1	4	0	2
223	8.58	1.92	0	1	4	0	1
224	15.98	3.00	1	0	4	0	3
225	13.42	1.58	0	1	4	0	2
226	16.27	2.50	1	1	4	0	2
227	10.09	2.00	1	1	4	0	2
228	20.45	3.00	0	0	5	1	4
229	13.28	2.72	0	0	5	1	2
230	22.12	2.88	1	1	5	1	2
231	24.01	2.00	0	1	5	1	4
232	15.69	3.00	0	1	5	1	3
233	11.61	3.39	0	0	5	1	2
234	10.77	1.47	0	0	5	1	2
235	15.53	3.00	0	1	5	1	2
236	10.07	1.25	0	0	5	1	2
237	12.60	1.00	0	1	5	1	2
238	32.83	1.17	0	1	5	1	2
239	35.83	4.67	1	0	5	1	3
240	29.03	5.92	0	0	5	1	3
241	27.18	2.00	1	1	5	1	2
242	22.67	2.00	0	1	5	1	2
243	17.82	1.75	0	0	5	1	2
244	18.78	3.00	1	0	3	1	2

CASE 19

RUBBERGATE

A congressional election is right around the corner. A small but vocal group of Republicans, hoping to unseat Democratic incumbents in the upcoming election, have received a significant amount of attention from the media. They charge ongoing fiscal irresponsibility on the part of the current members of Congress, and their accusations have struck a chord with the American public. The Republicans are using the House Banking Scandal as an example of the type of fiscal improprieties that are "deeply ingrained in the minds of our current elected officials." The House Banking Scandal, also dubbed "Rubbergate," involved the discovery that certain members of the House of Representatives bounced numerous bank checks with impunity in the early 1990s. At the time, Rubbergate received significant press, as, for example, newspapers across the country listed the names and numbers of checks bounced by their representatives. Now the Republican hopefuls are reviving the Rubbergate scandal.

Frank works as a statistician for the Republican party. He has been asked to study the earlier Rubbergate scandal for possible political ammunition. He has collected a random sample of 146 members of the 1992 House of Representatives, which may be found in the file named RUBBERG8. Eight characteristics were collected about each representative. Whether the representative wrote bad checks is indicated by the variable named CHECKS. This variable is coded 1 for a bad-check writer and 0 otherwise. PARTY is the variable indicating political affiliation. Democrats are coded with 1, and Republicans are coded 0. There weren't any Independents in this random sample. SERVICE is the total number of years of service in the House. Service beginning in or after November in a given year is not counted as a year of service in this data set. The representative's age in years is recorded in the variable named AGE. VOTES measures

---

This case is based on data provided to us by Chris A. Wilson. The data come from three sources: *Politics in America,* Congressional Quarterly, Inc., Washington, D.C., 1991, pp. 9–12; *Vital Statistics on Congress 1991–1992,* Congressional Quarterly, Inc., Washington, D.C., 1992, pp. 214–31; and *Rocky Mountain News,* "The House List," Friday, April 17, 1992, pp. A50–A51.

the returns (in percentages) for the most recent election in which that member participated. DISTRICT measures the representative's home district population rounded to the nearest thousand. STATE is like DISTRICT except it is the population of the representative's home state. If the representative wrote bad checks, Frank also recorded how many he or she wrote. This information is in the variable called BADCHKS.

Frank wants to prepare a report that characterizes the data set. In particular, what should he say about the following issues?

1 What is the general nature of the sample in terms of average and typical fluctuation around the average for the variables named SERVICE, AGE, VOTES, DISTRICT, and STATE?

2 How do these variables differ in a comparison of check-bouncers and non-check-bouncers?

3 How does party affiliation enter into the analysis?

4 For those individuals who did write bad checks, summarize the total number of bad checks written in some appropriate fashion.

DATA SET

CHECKS	PARTY	SERVICE	AGE	VOTES	DISTRICT	STATE	BADCHKS
1	0	6	58	99.6	594	4040	9
1	1	24	69	99.7	574	4040	4
0	1	8	52	92.8	537	4040	0
1	0	4	47	99.5	829	3665	32
0	0	4	48	61.3	747	3665	0
1	1	6	62	60.4	613	2350	1
1	0	0	40	43.0	651	29760	3
0	1	12	48	54.7	758	29760	0
1	1	16	45	60.5	666	29760	99
1	1	28	75	62.7	692	29760	13
1	1	16	59	58.0	566	29760	3
1	1	14	52	74.2	653	29760	12
1	0	17	64	54.6	647	29760	3
0	0	18	68	60.0	638	29760	0
1	1	28	74	70.0	601	29760	11
0	1	12	56	72.7	600	29760	0
1	1	10	64	67.1	599	29760	1
0	1	8	60	60.7	598	29760	0
0	0	8	63	49.7	980	29760	0
0	0	2	38	67.6	691	29760	0
1	0	8	59	68.1	841	29760	4
1	1	18	50	63.7	453	3294	5
0	0	0	47	54.1	529	3294	0
1	1	9	54	71.4	541	3287	60
1	0	3	45	76.5	519	3287	18
1	1	8	43	65.5	666	666	3
1	1	42	80	72.7	569	12938	4
1	0	2	46	59.2	765	12938	9

(continues on the following page)

CHECKS	PARTY	SERVICE	AGE	VOTES	DISTRICT	STATE	BADCHKS
0	0	8	60	58.1	769	12938	0
1	0	8	66	100.0	766	12938	8
0	0	10	51	97.8	544	12938	0
0	0	2	38	60.4	634	12938	0
1	1	10	51	73.0	557	6478	819
1	1	4	50	75.6	540	6478	125
0	1	8	64	68.7	563	6478	0
0	1	14	57	55.8	777	6478	0
0	1	13	63	66.3	594	1108	0
1	1	8	72	93.8	413	11431	716
0	1	2	59	59.2	533	11431	0
1	1	18	59	79.9	445	11431	18
1	0	11	55	67.7	566	11431	1
0	0	6	61	65.8	611	11431	0
1	1	8	39	86.8	467	11431	9
1	1	8	46	66.2	490	11431	12
0	1	6	41	66.0	493	5544	0
1	1	2	38	60.7	573	5544	21
1	0	24	63	57.6	562	5544	61
1	1	24	58	66.4	510	5544	1
1	1	4	47	99.2	462	2777	4
0	0	4	42	71.8	449	2777	0
0	0	6	62	60.1	549	2478	0
1	1	16	53	86.9	527	3685	152
0	0	4	59	69.3	563	3685	0
1	1	6	36	50.8	492	3685	514
1	1	14	43	52.5	467	4212	8
0	1	0	37	60.1	637	1228	0
1	0	12	43	51.0	591	1228	1
0	1	4	47	69.7	541	4782	0
1	1	12	58	65.3	670	4782	6
0	0	32	69	77.5	538	6016	0
0	1	10	50	65.5	545	6016	0
1	1	14	44	99.9	525	6016	92
1	1	18	53	53.4	593	6016	10
1	0	14	58	64.1	536	9295	17
1	0	6	48	75.4	579	9295	20
1	1	17	59	68.6	489	9295	201
1	0	12	58	61.3	521	9295	878
1	1	10	42	63.6	502	9295	547
0	1	8	59	69.7	485	9295	0
1	0	10	38	61.8	480	4375	125
0	1	12	52	72.9	498	4375	0
1	1	16	56	72.9	518	4375	2
0	1	24	70	100.0	530	2573	0
1	1	22	59	60.9	492	5117	328

(continues on the following page)

CASE 19 RUBBERGATE

CHECKS	PARTY	SERVICE	AGE	VOTES	DISTRICT	STATE	BADCHKS
1	1	14	59	61.8	615	5117	9
0	0	2	61	52.1	610	5117	0
1	1	12	53	61.1	417	799	66
0	1	2	49	57.9	560	1578	0
1	0	8	69	63.5	609	1202	2
1	1	0	33	54.3	567	7730	1
0	0	10	37	62.9	615	7730	0
1	0	18	59	74.6	520	7730	8
1	1	2	56	81.5	465	7730	6
1	0	6	47	58.1	613	7730	2
0	0	10	63	100.0	497	7730	0
1	1	16	41	55.8	567	7730	151
1	0	10	48	54.6	513	17991	4
1	1	24	70	72.3	527	17991	133
1	1	8	56	92.9	541	17991	408
1	0	1	32	60.0	536	17991	5
1	1	14	63	80.4	534	17991	3
1	1	2	53	62.8	517	17991	1
1	1	2	43	64.1	520	17991	15
1	0	10	46	100.0	545	17991	8
1	0	28	71	63.0	531	17991	3
0	1	16	51	55.0	499	17991	0
1	1	25	77	64.8	587	6629	63
1	1	4	50	58.1	699	6629	8
0	1	18	51	65.6	601	6629	0
0	0	4	64	61.8	572	6629	0
0	1	0	39	51.1	509	10847	0
1	0	10	46	61.7	511	10847	6
0	0	0	54	62.1	527	10847	0
0	0	24	73	63.2	527	10847	0
0	1	14	59	56.7	521	10847	0
1	0	18	66	58.9	511	10847	14
1	1	8	43	64.8	517	10847	397
0	0	4	56	56.0	518	3146	0
1	1	10	40	73.6	553	3146	8
1	1	16	48	63.1	617	2842	83
0	1	4	43	85.8	539	2842	0
1	0	16	61	57.1	581	11882	4
0	1	22	63	57.0	537	11882	0
1	1	12	44	56.6	578	11882	50
1	1	6	53	100.0	507	11882	7
0	1	10	54	71.8	453	11882	0
0	0	8	60	100.0	534	11882	0
1	1	22	64	65.6	469	11882	3
0	0	12	61	59.4	503	11882	0
0	0	4	63	65.5	616	3487	0

(concludes on the following page)

CHECKS	PARTY	SERVICE	AGE	VOTES	DISTRICT	STATE	BADCHKS
1	1	4	51	61.4	577	3487	2
0	1	4	44	67.6	696	696	0
1	1	16	61	53.0	521	4877	8
1	1	6	41	66.7	607	4877	6
1	1	16	45	58.1	462	4877	388
1	0	8	43	99.6	714	16987	18
0	0	6	41	66.5	705	16987	0
0	1	38	68	57.7	564	16987	0
1	1	2	38	71.3	609	16987	3
1	1	26	63	100.0	658	16987	284
1	1	1	49	99.6	450	16987	3
1	0	4	43	74.8	670	16987	1
0	1	12	48	100.0	635	16987	0
1	1	8	53	100.0	594	16987	18
0	1	0	41	58.3	604	1723	0
1	1	4	60	75.0	654	6187	1
0	1	3	45	99.4	544	6187	0
1	1	0	45	51.7	747	6187	3
1	0	6	52	52.0	657	4867	58
0	0	10	57	70.7	562	4867	0
0	1	2	53	72.3	558	4867	0
0	1	8	39	55.5	488	1794	0
1	1	20	52	99.4	533	4892	6
1	1	7	47	69.2	537	4892	1
1	1	22	52	99.5	544	4892	64
0	0	2	54	55.1	454	454	0

CASE 20

HITECH ENGINEERING

HiTech Engineering, Inc. (HEI), is a company located in Minnesota that designs and manufactures a variety of industrial products. HEI sells its products directly to existing customers and also uses authorized distributors around the globe to assist in sales and distribution. Although HEI has an edge on the market in terms of engineering and manufacturing expertise, it falls a bit short in knowing the best way to market and sell its latest industrial inventions and innovations.

HEI gets word to the market about a new product in various ways. For instance, sometimes they purchase magazine advertisements and sometimes they issue news releases. Because HEI produces industrial products and not consumer products, it never uses electronic media such as television or radio, but concentrates on print media for the purpose of advertising.

HEI management would like to understand how effective each of the five basic types of advertising they've used has been.

1 Magazines. Advertisements are placed in about 150 different industrial magazines and trade journals. These ads are designed by HEI's advertising department.

2 Postcard decks. Publishers of various magazines and journals assemble postcard announcements of new products from a number of different companies. The postcard decks are then sent out to its list of subscribers. HEI must pay for the postcard deck service.

3 Editorials. HEI's advertising department occasionally produces editorials for magazines or trade publications and hopes that the editorial will be published. In response to the editorial, readers will occasionally write back to HEI requesting more

---

These data are real even though the name of the company and its location have been changed. Tessa E. Alexander and Edwin C. Peterson have granted us permission to use these data, which they originally collected for study.

information. HEI's expenses for this type of marketing are employee salaries for writing the editorial and preparing a camera-ready version of the editorial.

4 News releases. A news release is a one paragraph description of a new product that typically appears in the back of a magazine. News releases will be published for free by magazines that are already carrying HEI advertisements as a courtesy to their clients. Like requests for literature (below), interested readers send a postcard to the magazine, which is then forwarded to HEI. Costs incurred for this type of advertising are employee salaries for processing requests.

5 Literature. Readers of magazines or trade journals may make a request for literature or more information about HEI by circling a request number on a "bingo card" and sending it back to the magazine or journal. Requests are then sent from the magazine or journal to HEI, which responds to the requests appropriately.

Requests for information by prospective customers are called "leads." By asking customers where they heard about the new product, the advertising group was able to compile information about the number of leads generated by each of these media types; they also know the expenses incurred in each media class. The data cover the last 17 months. These data have been assembled in the file named LEADATA.

The data set contains the following information. MONTH indicates the appropriate month, which takes on values from 1 to 17. RESPONSE is the number of responses for information about the company's product. SPEND indicates expenses incurred in that month for that media type in dollars. Finally, MAGS, POST, EDIT, NEWS, and LIT are binary variables indicating which media type resides in that row of the data set. These correspond to the five media described above.

What is the most effective type of advertising outlet for HEI? Do you have any recommendations on the appropriate mix of the five different types of advertising?

DATA SET

MONTH	RESPONSE	SPEND	MAGS	POST	EDIT	NEWS	LIT
1	150	2505	1	0	0	0	0
2	76	1760	1	0	0	0	0
3	79	3225	1	0	0	0	0
4	82	7527	1	0	0	0	0
5	98	7100	1	0	0	0	0
6	106	6000	1	0	0	0	0
7	81	9330	1	0	0	0	0
8	87	3910	1	0	0	0	0
9	64	7878	1	0	0	0	0
10	111	4233	1	0	0	0	0
11	102	2475	1	0	0	0	0
12	65	1800	1	0	0	0	0
13	42	1689	1	0	0	0	0
14	30	1559	1	0	0	0	0
15	40	865	1	0	0	0	0
16	42	0	1	0	0	0	0
17	10	2635	1	0	0	0	0
1	656	9490	0	1	0	0	0

(continues on the following page)

MONTH	RESPONSE	SPEND	MAGS	POST	EDIT	NEWS	LIT
2	966	10660	0	1	0	0	0
3	503	0	0	1	0	0	0
4	230	5534	0	1	0	0	0
5	379	9517	0	1	0	0	0
6	690	10852	0	1	0	0	0
7	560	6321	0	1	0	0	0
8	498	7263	0	1	0	0	0
9	459	5829	0	1	0	0	0
10	315	4058	0	1	0	0	0
11	226	0	0	1	0	0	0
12	57	0	0	1	0	0	0
13	160	7131	0	1	0	0	0
14	642	3773	0	1	0	0	0
15	153	1436	0	1	0	0	0
16	421	2377	0	1	0	0	0
17	666	4373	0	1	0	0	0
1	60	1779	0	0	1	0	0
2	29	2100	0	0	1	0	0
3	64	1899	0	0	1	0	0
4	105	2200	0	0	1	0	0
5	69	1890	0	0	1	0	0
6	21	1900	0	0	1	0	0
7	24	2112	0	0	1	0	0
8	17	2000	0	0	1	0	0
9	10	1877	0	0	1	0	0
10	50	2200	0	0	1	0	0
11	54	1788	0	0	1	0	0
12	42	2300	0	0	1	0	0
13	56	2166	0	0	1	0	0
14	102	1877	0	0	1	0	0
15	37	1944	0	0	1	0	0
16	36	2000	0	0	1	0	0
17	6	2100	0	0	1	0	0
1	81	189	0	0	0	1	0
2	139	176	0	0	0	1	0
3	174	231	0	0	0	1	0
4	152	179	0	0	0	1	0
5	166	215	0	0	0	1	0
6	98	197	0	0	0	1	0
7	82	241	0	0	0	1	0
8	66	187	0	0	0	1	0
9	74	185	0	0	0	1	0
10	42	216	0	0	0	1	0
11	53	197	0	0	0	1	0
12	26	193	0	0	0	1	0
13	71	216	0	0	0	1	0
14	74	195	0	0	0	1	0

(concludes on the following page)

MONTH	RESPONSE	SPEND	MAGS	POST	EDIT	NEWS	LIT
15	82	208	0	0	0	1	0
16	138	185	0	0	0	1	0
17	91	194	0	0	0	1	0
1	2	375	0	0	0	0	1
2	34	405	0	0	0	0	1
3	121	413	0	0	0	0	1
4	144	378	0	0	0	0	1
5	78	389	0	0	0	0	1
6	116	385	0	0	0	0	1
7	220	407	0	0	0	0	1
8	29	407	0	0	0	0	1
9	24	403	0	0	0	0	1
10	15	399	0	0	0	0	1
11	263	408	0	0	0	0	1
12	117	413	0	0	0	0	1
13	96	401	0	0	0	0	1
14	31	411	0	0	0	0	1
15	84	400	0	0	0	0	1
16	104	420	0	0	0	0	1
17	31	410	0	0	0	0	1

CASE 21

401(K)

Best Retirement, Inc. (BRI), sells retirement plans to corporations around the country. To capture a market niche, BRI has decided to target smaller firms—those with 500 or fewer employees. BRI is positioned to offer a variety of retirement plans for its clients, from 401(k) plans to individual retirement accounts. BRI also writes group health and group life insurance policies, although the major portion of their revenue comes from retirement packages.

It is extremely important that BRI be able to estimate the dollar amount contributed to each plan by year's end. It would be nice to be able to estimate total contributions before the plan is even written, so that BRI can make internal revenue and cost projections. For this reason, BRI has hired an underwriter who estimates year-end contributions based on information provided by the client on an application form. In the past, the underwriter estimates have not been very accurate. Your task is to prepare more accurate estimates of annual contributions for these retirement plans.

To this end, BRI has provided you with lots of data about 401(k) retirement plans written by BRI representatives from January to March of last year. The file, named 401(k), contains 92 plans sold over this time period with the ultimate contributions to each plan at year's end. This variable, labeled CONTRIB, is the one you are attempting to understand. The underwriter's estimate (ESTIMATE) is also given; this is the variable you would like to improve upon.

A number of factors are likely to influence employee contributions to a retirement plan. Clearly employees' salaries would be an important factor. "Matching-contribution" provisions would also be a potential determinant; we might expect that employer contributions matching employee contributions would induce greater employee participation

---

These data are real, but the names of the company and of the person who supplied the data to us have been withheld upon their request.

in the retirement plan. Firms with high employee turnover rates might have low employee participation rates, all else the same. BRI also suspects that it is to their advantage to offer "package deals" and that 401(k) contributions are more likely if BRI also writes group health and group life insurance policies for that client.

Several other provisions of the policy might influence contribution levels. For instance, some plans have fail-safe provisions. This means that if the plan fails to meet government antidiscrimination rules, the employer will make contributions to the lower-paid employees' accounts. Also, a plan might have immediate vesting of employer contributions; employees who withdraw their money also receive all of the employer's matching contributions at the time of withdrawal if the plan is vested.

Finally, BRI has one sales representative, Susan Shepard, who has been specifically trained to deal exclusively with 401(k) plans. Other sales representatives can write 401(k) plans, but don't have quite the in-depth knowledge that Ms. Shepard does. BRI would like to know if Susan's expertise is a factor that influences year-end contributions to 401(k) plans. If so, BRI would like to consider training some of its other sales representatives.

In the data set shown below, CONTRIB is the variable indicating contribution to the plan in dollars at the end of its first year. SUSAN is a categorical variable taking on the value of 1 if Susan Shepard sold the policy and 0 if she did not. GROUP takes on the value of 1 if the client also has a group life or group health insurance policy with Best Retirement; if not, GROUP is 0. TURNOVER is the employee turnover rate. It measures the percentage of employees terminating employment each year. ESTIMATE is the estimated end-of-year contributions in dollars and SALARY is the average annual employee salary in dollars. ELIGIBLE is the number of employees eligible to participate in the 401(k) plan. VEST and FAILSAFE are both binary variables: does the plan have immediate vesting of employer contributions (0 = no, 1 = yes) and does the plan have a fail-safe provision (0 = no, 1 = yes)? MATCH is the percentage of the contributions matched by the employer.

Can you do better than the underwriter at estimating year-end contributions?

DATA SET

CONTRIB	GROUP	TURNOVER	ELIGIBLE	VEST	FAILSAFE	MATCH	SALARY	ESTIMATE	SUSAN
36675	1	14.00	69	1	0	25	26296.3	75432	0
63733	0	10.00	33	1	0	50	14133.7	50000	0
25560	1	8.00	21	1	0	50	24000.0	45000	0
177970	0	10.00	67	1	0	25	36833.3	235000	0
86873	1	10.00	47	1	0	50	41140.1	146965	0
39051	0	10.00	12	1	0	0	66463.8	95000	0
131449	1	12.00	85	1	0	25	25929.3	155045	0
30711	0	10.00	40	1	0	50	17198.7	54392	1
13691	0	10.00	30	1	1	0	28788.6	50000	0
49587	1	10.00	63	1	0	25	32124.4	100000	0
37898	0	10.00	25	1	1	100	35288.1	86000	0
41686	1	0.10	30	0	0	100	23266.7	56000	0
107657	0	14.00	21	1	0	50	54798.0	115650	0
39811	0	14.00	29	1	0	50	18617.9	50000	0
148274	0	14.00	99	1	0	0	25516.6	252316	0

(continues on the following page)

CONTRIB	GROUP	ELIGIBLE TURNOVER	FAILSAFE VEST	MATCH	SALARY	ESTIMATE	SUSAN
91496	0	10.80	21	1	1	25	41250.0	113152	1
126988	0	10.00	28	1	0	0	21599.3	77280	0
158941	0	7.50	283	1	0	25	20405.2	321750	0
13691	0	12.00	50	0	0	100	23778.9	35772	0
118396	0	14.00	250	1	0	0	28626.9	243750	0
102670	0	7.00	30	1	0	0	37906.4	101431	0
21012	0	12.00	28	1	0	100	33941.2	78330	0
58448	0	14.00	36	1	0	25	23492.0	60304	0
39787	0	12.00	23	1	0	50	28222.1	41458	0
73614	0	14.00	66	0	0	0	35874.8	144316	0
218158	0	12.00	150	0	0	100	21626.1	200000	1
41335	0	12.00	121	1	0	20	57182.2	141481	0
19120	1	14.00	69	1	0	0	21503.3	103171	0
73053	1	11.00	175	1	0	0	25102.2	152646	0
166110	1	14.00	146	1	0	25	37219.7	267268	0
141009	1	14.00	70	1	0	0	20380.5	154135	0
98052	0	13.00	176	1	0	0	34764.0	276222	0
42293	1	14.00	38	0	0	0	34182.8	90074	0
169983	0	12.00	144	1	0	25	26809.3	255126	0
71849	0	14.00	69	0	0	0	67391.1	119856	0
28429	1	12.00	110	1	0	0	30397.0	144000	0
24154	1	14.00	21	1	0	75	33374.8	50088	0
12038	1	14.00	49	1	0	0	21133.9	51592	0
289886	1	11.80	391	1	0	0	27309.8	432562	0
150329	1	11.00	80	0	1	0	37927.6	173846	0
69020	0	14.00	32	1	0	0	43603.4	90000	1
25749	1	18.00	281	1	0	0	18943.7	120000	0
32147	1	14.00	63	1	0	0	30144.4	103320	0
101665	1	11.80	45	0	0	25	35661.8	126625	0
224721	0	8.21	437	1	0	25	24986.9	679972	0
249063	1	14.00	400	1	0	25	20511.8	425040	0
17865	0	12.00	9	1	0	100	59528.6	40500	0
28912	0	14.00	31	1	0	50	27289.5	59271	0
42703	0	14.00	25	1	0	0	34137.7	51026	0
120940	0	14.00	50	1	0	50	35888.3	106000	0
149654	0	14.00	230	1	0	40	26807.3	200000	0
277688	0	12.00	244	1	0	0	36729.4	263676	1
25214	1	12.00	54	1	0	10	18039.7	73031	0
141359	0	12.00	234	0	0	0	29970.1	217420	0
57282	1	12.00	23	1	0	100	26647.8	50000	0
76269	0	12.00	23	1	0	25	37104.9	64488	0
75572	0	12.00	63	1	0	10	42790.9	175000	0
71330	1	13.00	85	1	0	50	25932.4	192000	0
28748	1	14.00	69	1	0	100	21250.7	123084	0
151405	1	14.00	141	1	0	25	30506.1	262005	0
24285	0	14.00	49	1	1	10	15423.5	46104	1
58131	1	14.00	130	0	0	0	38793.3	140576	0

(concludes on the next page)

CASE 21 401(K)

CONTRIB	GROUP	TURNOVER	ELIGIBLE	VEST	FAILSAFE	MATCH	SALARY	ESTIMATE	SUSAN
44323	1	14.00	41	1	0	0	37380.2	97680	0
318022	0	14.00	154	0	0	100	51865.4	524000	0
34052	0	14.00	17	0	0	100	30398.9	61222	0
82837	0	12.00	28	1	0	25	41309.2	70275	0
86394	0	14.00	100	1	0	0	30265.6	264075	1
26810	1	14.00	126	1	0	0	29346.0	281954	0
37516	1	12.00	62	1	0	50	24814.9	65052	0
44228	1	14.00	135	1	0	25	19549.9	165102	0
24541	1	12.00	60	1	0	0	39189.3	121135	0
80602	1	14.00	96	1	0	50	26440.6	165240	0
56209	1	14.00	64	1	0	0	45428.3	170928	0
187159	1	12.50	350	1	0	0	23659.6	394980	0
34878	1	14.00	60	1	0	25	32408.3	135936	0
126924	1	14.00	86	1	0	0	37579.8	150000	0
197015	0	14.00	27	1	0	50	55804.3	123405	1
228169	0	14.00	84	0	0	20	51481.8	399431	1
147450	1	13.00	176	1	0	25	25424.2	251064	0
39691	1	14.00	68	0	0	25	27399.6	134133	0
52618	0	14.00	37	1	0	20	36806.7	113010	0
25530	1	14.00	37	1	0	0	43253.4	87000	0
39311	1	15.00	70	1	0	0	22034.8	72680	0
45969	1	14.00	75	1	0	25	18465.4	82320	0
13604	1	14.00	36	1	0	0	30352.5	50930	0
304864	0	14.00	141	1	1	25	42847.9	271153	0
66685	0	12.00	29	1	0	0	47012.6	118145	0
26192	1	20.00	49	1	0	25	2234.4	67340	0
67353	0	14.00	52	1	0	100	37295.0	220153	1
22954	0	14.00	150	1	0	25	18352.1	147000	0
109116	1	14.00	246	0	0	0	31887.7	506530	0
18022	0	14.00	51	1	0	25	71320.0	98076	0

CASE **22**

LATE CHARTS

Organizations that serve their customers by appointments at the company's offices must have some mechanism for ensuring that appropriate customer records are in the right office in time for the appointment. Misplaced records cause delays and hard feelings; in extreme cases, appointments are canceled for lack of appropriate records. Small organizations may have a simple "system" to provide records: the agent assigned may retrieve the file from a cabinet. If the file isn't there, she may look around the office until she finds it. In larger organizations, though, the large number of records involved usually means that some more formal arrangement is needed to make records available to the appropriate agent for the appointment. The records delivery system (if we may call it that) is often connected to the system that schedules the appointments. Tax audits, counseling sessions at social agencies, and insurance planning sessions share this need, as do many medical offices.

At a large medical center in the southwest, patients fill out forms and doctors dictate reports that are kept in paper form in file folders in a central filing room in the basement of the building. The center has considered converting medical records to a computerized form, but the conversion process is remarkably complicated when all its details and implications are considered, so while such a conversion is an active project, it is not yet complete.

Patients may schedule appointments with any of the more than 60 physicians either personally or by telephone. In either case, an operator consults a computer, which displays available appointment times and any restrictions on the kinds of appointments for which they may be scheduled. The operator works with the patient to find a mutually agreeable time and indicates that time to the computer system, which updates the reservation database. If the patient is physically present when the reservation is made, the

---

This case is an amalgam of several situations with which the authors have dealt.

operator gives him or her a card with the scheduled time on it; otherwise, the operator confirms the time orally with the patient.

Each night, the computer produces a list of patients who are scheduled for the following day for each physician and a corresponding list of charts to be pulled from the central filing area and delivered to the physicians' offices. A medical records clerk works in the evening to be sure the files are delivered. First, he collects and re-files any records returned from that day's appointments or from the centralized dictation area. Then he pulls the files needed for the next day and delivers them along with the list of appointments for each physician to each physician's office.

Occasionally, charts are unavailable. Usually this happens because they are still in use by a physician who saw the patient recently, but occasionally they are misplaced, left in someone's desk drawer or waiting in some other pile of folders somewhere in the building. The clerk enters the patient names for such files on a separate list and leaves the list for the first-shift records staff. They attack the problem as their first order of business in the morning. Usually, the computerized records of which physician last saw the patient provide a clue they can use to track down the file, but in perhaps 10 or 15 percent of the cases, they can't deliver the file in time for the scheduled appointment.

As part of a quality management initiative at the medical center, the medical records staff sampled the appointment records to assess the frequency of late charts. They sampled 100 patient appointments at random for each of 20 successive weeks. By consulting the physicians' staffs, they determined the number of appointments for which patient charts had been delivered late. The records staff knew that "late chart" was to some extent a subjective judgment on the part of the nurses and receptionists in the physicians' offices, since they kept no record or time-stamp of when the chart was delivered. Still, they figured, it was their (internal) customer's impression of quality, and it mattered. The data they collected are given below (and in file CHARTS). WEEK is the week number (1 through 20); NLATE is the number of late charts.

What do you conclude about the chart delivery process?

DATA SET

WEEK	NLATE	WEEK	NLATE	WEEK	NLATE	WEEK	NLATE
1	14	6	11	11	8	16	10
2	10	7	10	12	12	17	8
3	12	8	12	13	9	18	12
4	13	9	13	14	10	19	10
5	9	10	10	15	11	20	16

CASE 23

MONEY SUPPLY AND INTEREST RATES

As Kellie sat in her economics class, she wondered whether her professor spent too much time in the ivory tower. She thought her question to the professor was quite straightforward, but he seemed flustered and uncertain in his response.

It all started innocently enough. The lecture was about monetary policy and the methods by which the Federal Reserve can control monetary instruments. At some point during the hour, Professor Hicks said, "When the Feds increase the money supply, by printing more money or reducing the reserve requirement, for example, the interest rate will increase." He then went into a long discussion of the mechanisms that cause this positive relationship between money and interest rates.

Although it was all very interesting, Kellie thought the ideas were too nebulous to be of any practical value. So she raised her hand. "Professor Hicks, I certainly enjoy your lively lectures, but there is one point on which I am very confused. What do you mean when you say *the* interest rate and *the* money supply? For instance, last week you told us about three different measures of money supply—M1, the narrow measure; M2, an intermediate quantity; and M3, the broadest measure of money. Are you referring to all three of these measures when you say *the* money supply?"

Professor Hicks replied that he meant the money supply in a generic sense and that the general theory applied equally to all three measures.

"Well, then, Professor Hicks, what is *the* interest rate? I mean, I have an interest rate on my new car loan, my parents have an interest rate on their mortgage, my credit card has an interest rate, and they're all different. Plus, last week, you were talking about the prime rate, the discount rate, a commercial paper rate, and a whole bunch of other things. Are you referring to only one of these interest rates, or all of them?"

Professor Hicks said, "You know, you make an interesting point. We would expect the different interest rates to react somewhat differently to changes in the stock of money, but I haven't really spent much time thinking about the issue you raise. Give me time to collect some data, and I'll report back to you during our next class meeting."

CASE 23 MONEY SUPPLY AND INTEREST RATES

The file named MONEY contains a variety of interest rate and money stock variables measured on a monthly basis beginning in January 1980. These data are not seasonally adjusted. Asterisks at the end of the data indicate that the data were not available at the time Professor Hicks's student assistant prepared the data set for him. Updated monthly observations are available on the Federal Reserve Board's website: the money data from *www.bog.frb.us/Releases/H6/hist/* and the interest rate data from *www.bog.frb.fed.us/Releases/H15/data.HTM#FN3*.

Are some of the interest rates more responsive to certain measures of money supply than others? Is there any evidence to suggest that certain interest rates are unrelated to money supply? Professor Hicks wants to be fully prepared for the next class.

DEFINITIONS

M1	Currency in circulation, travelers checks, demand deposits, and other checkable deposits, in billions of dollars
M2	M1 plus retail money market mutual funds, savings accounts, and small time deposits, in billions of dollars
M3	M2 plus large time deposits, Eurodollars, and institutions-only money market mutual funds, in billions of dollars
PRIME	The interest rate banks charge their best customers for short-term loans
DISCOUNT	The interest rate charged on loans made by the New York Federal Reserve Bank to its member banks
HOME	The interest rate on 30-year, fixed-rate, conventional mortgage loans
T-BILL	Yield on three-month U.S. Treasury Bills, new issues

DATA SET

M1	M2	M3	PRIME	DISCOUNT	HOME	T-BILL
390.5	1487.5	1822.8	15.25	12.00	12.88	12.04
380.7	1487.4	1831.7	15.63	12.52	13.04	12.82
382.2	1498.4	1844.7	18.31	13.00	15.28	15.53
386.9	1509.1	1853.2	19.77	13.00	16.33	14.00
377.6	1505.2	1855.0	16.57	12.94	14.26	9.15
387.4	1528.7	1876.6	12.63	11.40	12.71	7.00
394.6	1550.3	1902.2	11.48	10.87	12.19	8.13
398.3	1561.9	1919.5	11.12	10.00	12.56	9.26
404.7	1574.3	1931.2	12.23	10.17	13.20	10.32
410.8	1589.8	1952.2	13.79	11.00	13.79	11.58
415.8	1600.4	1975.4	16.06	11.47	14.21	13.89
419.5	1606.2	1997.0	20.35	12.87	14.79	15.66
416.2	1611.8	2019.0	20.16	13.00	14.90	14.73
405.5	1611.9	2027.2	19.43	13.00	15.13	14.91
412.3	1634.5	2048.9	18.05	13.00	15.40	13.48
431.1	1665.8	2080.4	17.15	13.00	15.58	13.63
418.4	1656.8	2085.8	19.61	13.87	16.40	16.29
422.8	1668.5	2104.6	20.03	14.00	16.70	14.56

(continues on the following page)

M1	M2	M3	PRIME	DISCOUNT	HOME	T-BILL
427.6	1686.4	2131.7	20.39	14.00	16.83	14.70
425.9	1696.0	2153.4	20.50	14.00	17.29	15.61
427.0	1707.9	2172.0	20.08	14.00	18.16	14.95
429.6	1727.8	2195.9	18.45	14.00	18.45	13.87
435.0	1742.4	2218.5	16.84	13.03	17.83	11.27
447.0	1761.9	2245.1	15.75	12.10	16.92	10.93
448.5	1774.3	2265.8	15.75	12.00	17.40	12.41
432.4	1766.8	2263.9	16.56	12.00	17.60	13.78
435.6	1783.3	2286.2	16.50	12.00	17.16	12.49
451.1	1809.3	2316.4	16.50	12.00	16.89	12.82
440.9	1806.3	2319.5	16.50	12.00	16.68	12.15
446.3	1824.3	2338.8	16.50	12.00	16.70	12.11
449.4	1839.3	2356.8	16.26	11.81	16.82	11.92
449.8	1851.7	2383.6	14.39	10.68	16.27	9.01
456.2	1865.6	2398.2	13.50	10.00	15.43	8.20
465.8	1882.3	2423.1	12.52	9.68	14.61	7.75
474.4	1896.4	2438.0	11.85	9.35	13.83	8.04
485.8	1919.9	2450.4	11.50	8.73	13.62	8.02
482.8	1969.4	2479.3	11.16	8.50	13.25	7.81
474.1	1991.4	2495.3	10.98	8.50	13.04	8.13
482.7	2015.0	2518.0	10.50	8.50	12.80	8.30
498.5	2038.7	2543.5	10.50	8.50	12.78	8.25
493.9	2036.8	2547.9	10.50	8.50	12.63	8.19
503.5	2056.8	2569.2	10.50	8.50	12.87	8.82
510.5	2074.6	2585.7	10.50	8.50	13.42	9.12
508.3	2077.4	2598.8	10.89	8.50	13.81	9.39
511.3	2086.0	2614.4	11.00	8.50	13.73	9.05
517.1	2106.9	2637.2	11.00	8.50	13.54	8.71
521.9	2122.8	2667.3	11.00	8.50	13.44	8.71
533.2	2139.1	2695.5	11.00	8.50	13.42	8.96
530.2	2150.0	2709.1	11.00	8.50	13.37	7.14
516.9	2152.9	2720.9	11.00	8.50	13.23	9.03
523.3	2174.2	2752.7	11.21	8.50	13.39	9.08
539.0	2199.9	2786.9	11.93	8.87	13.65	9.69
530.7	2196.4	2804.2	12.39	9.00	13.94	9.90
541.4	2219.6	2832.5	12.60	9.00	14.42	9.94
543.4	2233.3	2857.8	13.00	9.00	14.67	10.13
539.0	2234.1	2872.5	13.00	9.00	14.47	10.49
542.5	2247.9	2892.2	12.97	9.00	14.35	10.41
542.2	2265.1	2920.5	12.58	9.00	14.13	9.97
549.8	2292.0	2954.7	11.77	8.83	13.64	8.79
564.6	2324.2	2992.9	11.06	8.37	13.18	8.16
561.1	2345.3	3014.0	10.61	8.00	13.08	7.76
551.9	2348.9	3019.7	10.50	8.00	12.92	8.17
558.4	2365.9	3039.9	10.50	8.00	13.17	8.57
575.1	2384.9	3051.1	10.50	8.00	13.20	8.00
569.4	2381.8	3056.2	10.31	7.81	12.91	7.56

(continues on the following page)

CASE 23 MONEY SUPPLY AND INTEREST RATES

M1	M2	M3	PRIME	DISCOUNT	HOME	T-BILL
585.2	2419.2	3090.2	9.78	7.50	12.22	7.01
592.0	2440.4	3102.8	9.50	7.50	12.03	7.05
595.0	2448.2	3120.2	9.50	7.50	12.19	7.18
602.0	2458.5	3138.3	9.50	7.50	12.19	7.08
605.4	2472.2	3158.2	9.50	7.50	12.14	7.17
615.0	2487.8	3181.5	9.50	7.50	11.78	7.20
633.4	2510.1	3212.2	9.50	7.50	11.26	7.07
626.7	2516.7	3232.4	9.50	7.50	10.88	7.04
612.9	2507.9	3231.5	9.50	7.50	10.71	7.03
624.4	2533.2	3262.3	9.10	7.10	10.08	6.59
647.1	2570.0	3299.0	8.83	6.83	9.94	6.06
645.8	2576.7	3307.0	8.50	6.50	10.14	6.12
663.0	2611.1	3336.4	8.50	6.50	10.68	6.21
673.5	2638.7	3370.6	8.16	6.16	10.51	5.84
678.5	2652.3	3396.2	7.90	5.82	10.20	5.57
684.7	2668.3	3421.3	7.50	5.50	10.01	5.19
692.3	2691.2	3442.0	7.50	5.50	9.97	5.18
708.9	2711.9	3463.8	7.50	5.50	9.70	5.35
740.0	2747.3	3501.4	7.50	5.50	9.31	5.49
737.4	2760.1	3523.3	7.50	5.50	9.20	5.45
717.3	2741.8	3509.4	7.50	5.50	9.08	5.59
723.3	2754.3	3522.3	7.50	5.50	9.04	5.56
752.2	2784.8	3551.8	7.75	5.50	9.83	5.76
739.5	2765.6	3550.1	8.14	5.50	10.60	5.75
744.0	2779.5	3573.1	8.25	5.50	10.54	5.69
746.3	2791.7	3585.9	8.25	5.50	10.28	5.78
744.3	2793.9	3608.0	8.25	5.50	10.33	6.00
744.7	2799.8	3628.7	8.70	5.95	10.89	6.32
753.3	2817.4	3653.2	9.07	6.00	11.26	6.40
755.7	2828.3	3673.8	8.78	6.00	10.65	5.81
765.5	2845.4	3685.8	8.75	6.00	10.65	5.80
764.4	2864.6	3708.9	8.75	6.00	10.43	5.90
744.7	2866.2	3714.8	8.51	6.00	9.89	5.69
751.8	2894.1	3746.3	8.50	6.00	9.93	5.69
778.2	2932.3	3785.0	8.50	6.00	10.20	5.92
763.7	2917.3	3786.4	8.84	6.00	10.46	6.27
778.8	2942.1	3815.6	9.00	6.00	10.46	6.50
785.7	2961.1	3839.3	9.29	6.00	10.43	6.73
781.2	2957.8	3849.8	9.84	6.37	10.60	7.02
779.9	2958.2	3858.0	10.00	6.50	10.48	7.23
781.0	2970.4	3876.9	10.00	6.50	10.30	7.34
787.1	2991.0	3906.3	10.05	6.50	10.27	7.68
803.3	3008.7	3925.3	10.50	6.50	10.61	8.09
792.3	3007.2	3927.8	10.50	6.50	10.73	8.29
771.8	2990.7	3918.4	10.93	6.59	10.65	8.48
774.8	3006.9	3946.9	11.50	7.00	11.03	8.83
790.4	3030.5	3965.3	11.50	7.00	11.05	8.70
766.4	3003.5	3945.0	11.50	7.00	10.77	8.40

(continues on the following page)

M1	M2	M3	PRIME	DISCOUNT	HOME	T-BILL
773.1	3030.0	3976.8	11.07	7.00	10.20	8.22
781.0	3063.1	4007.1	10.98	7.00	9.88	7.92
776.8	3078.8	4016.3	10.50	7.00	9.99	7.91
778.0	3094.1	4019.3	10.50	7.00	10.13	7.72
783.6	3117.5	4032.2	10.50	7.00	9.95	7.63
790.5	3143.6	4059.3	10.50	7.00	9.77	7.65
811.0	3173.0	4077.9	10.50	7.00	9.74	7.64
801.3	3177.9	4075.1	10.11	7.00	9.90	7.64
787.1	3177.1	4068.2	10.00	7.00	10.20	7.76
795.0	3199.6	4084.3	10.00	7.00	10.27	7.87
816.8	3226.9	4097.5	10.00	7.00	10.37	7.78
796.0	3193.3	4072.0	10.00	7.00	10.48	7.78
809.5	3217.1	4091.6	10.00	7.00	10.16	7.74
811.5	3231.4	4104.0	10.00	7.00	10.04	7.66
813.1	3245.6	4118.8	10.00	7.00	10.10	7.44
817.6	3255.4	4117.3	10.00	7.00	10.18	7.38
816.6	3260.2	4119.2	10.00	7.00	10.18	7.19
825.1	3272.6	4125.1	10.00	7.00	10.01	7.07
843.2	3291.7	4138.5	10.00	6.79	9.67	6.81
832.3	3296.4	4151.1	9.52	6.50	9.64	6.30
822.6	3303.3	4164.1	9.05	6.00	9.37	5.95
834.2	3334.0	4183.0	9.00	6.00	9.50	5.91
852.4	3356.0	4193.8	9.00	5.98	9.49	5.67
840.9	3335.6	4167.7	8.50	5.50	9.47	5.51
857.3	3355.2	4176.0	8.50	5.50	9.62	5.60
861.5	3361.1	4170.4	8.50	5.50	9.58	5.58
863.7	3357.3	4169.8	8.50	5.50	9.24	5.39
866.5	3352.8	4153.6	8.20	5.20	9.01	5.25
874.5	3359.0	4158.7	8.00	5.00	8.86	5.03
892.8	3376.4	4181.2	7.58	4.58	8.71	4.60
916.0	3393.2	4195.2	7.21	4.11	8.50	4.12
916.9	3390.3	4190.5	6.50	3.50	8.43	3.84
915.5	3396.6	4202.7	6.50	3.50	8.76	3.84
930.0	3412.8	4212.1	6.50	3.50	8.94	4.05
954.1	3426.8	4210.4	6.50	3.50	8.85	3.81
943.3	3395.9	4181.7	6.50	3.50	8.67	3.66
951.4	3398.0	4183.3	6.50	3.50	8.51	3.70
962.4	3402.4	4183.0	6.02	3.02	8.13	3.28
970.4	3405.1	4199.2	6.00	3.00	7.98	3.14
983.0	3410.2	4194.4	6.00	3.00	7.92	2.97
1001.2	3426.9	4196.0	6.00	3.00	8.09	2.84
1021.8	3440.4	4209.1	6.00	3.00	8.31	3.14
1046.0	3450.4	4207.8	6.00	3.00	8.22	3.25
1040.2	3433.8	4178.5	6.00	3.00	8.02	3.06
1022.3	3410.2	4165.9	6.00	3.00	7.68	2.95
1030.8	3420.7	4174.8	6.00	3.00	7.50	2.97
1057.8	3441.6	4194.8	6.00	3.00	7.47	2.89
1057.2	3435.7	4198.4	6.00	3.00	7.47	2.96

(continues on the following page)

M1	M2	M3	PRIME	DISCOUNT	HOME	T-BILL
1072.2	3450.2	4206.1	6.00	3.00	7.42	3.10
1083.6	3453.8	4203.2	6.00	3.00	7.21	3.05
1088.6	3453.1	4207.7	6.00	3.00	7.11	3.05
1099.2	3455.8	4209.6	6.00	3.00	6.92	2.96
1112.0	3464.5	4225.3	6.00	3.00	6.83	3.04
1129.5	3486.2	4254.7	6.00	3.00	7.16	3.12
1153.7	3506.6	4274.8	6.00	3.00	7.17	3.08
1142.2	3494.2	4263.6	6.00	3.00	7.06	3.02
1124.1	3475.1	4231.6	6.00	3.00	7.15	3.21
1131.4	3492.8	4248.5	6.06	3.00	7.68	3.52
1152.5	3520.7	4275.2	6.45	3.00	8.32	3.74
1132.4	3490.9	4252.0	6.99	3.24	8.60	4.19
1141.9	3491.7	4262.1	7.25	3.50	8.40	4.18
1150.7	3504.7	4282.3	7.25	3.50	8.61	4.39
1144.1	3495.7	4282.6	7.51	3.76	8.51	4.50
1146.6	3491.3	4283.7	7.75	4.00	8.64	4.64
1148.0	3492.9	4298.2	7.75	4.00	8.93	4.96
1156.1	3506.1	4323.3	8.15	4.40	9.17	5.25
1174.4	3522.5	4347.4	8.50	4.75	9.20	5.64
1159.3	3510.2	4353.8	8.50	4.75	9.15	5.81
1135.1	3489.7	4342.4	9.00	5.25	8.83	5.80
1139.1	3506.6	4369.7	9.00	5.25	8.46	5.73
1160.1	3536.9	4407.1	9.00	5.25	8.32	5.67
1133.8	3520.9	4412.0	9.00	5.25	7.96	5.70
1140.8	3560.8	4460.1	9.00	5.25	7.57	5.50
1145.6	3586.6	4491.7	8.80	5.25	7.61	5.47
1139.3	3603.6	4524.4	8.75	5.25	7.86	5.41
1138.5	3613.5	4539.3	8.75	5.25	7.64	5.26
1132.9	3621.5	4561.5	8.75	5.25	7.48	5.30
1138.7	3643.2	4587.1	8.75	5.25	7.38	5.35
1152.8	3675.3	4612.0	8.65	5.25	7.20	5.16
1130.1	3673.3	4626.5	8.50	5.24	7.03	5.02
1105.7	3668.6	4641.0	8.25	5.00	7.08	4.87
1117.8	3715.7	4691.9	8.25	5.00	7.62	4.96
1131.3	3741.6	4715.7	8.25	5.00	7.93	4.99
1105.3	3709.1	4712.7	8.25	5.00	8.07	5.02
1114.2	3739.2	4748.7	8.25	5.00	8.32	5.11
1109.8	3753.1	4767.5	8.25	5.00	8.25	5.17
1096.5	3761.0	4788.1	8.25	5.00	8.00	5.09
1090.2	3762.5	4805.4	8.25	5.00	8.23	5.15
1076.9	3769.4	4845.4	8.25	5.00	7.92	5.01
1085.3	3801.2	4884.2	8.25	5.00	7.62	5.03
1103.1	3837.7	4935.0	8.25	5.00	7.60	4.87
1085.9	3836.3	4945.6	8.25	5.00	7.82	5.05
1066.5	3834.0	4968.2	8.25	5.00	7.65	5.00
1067.4	3870.0	5015.2	8.30	5.00	7.90	5.14
1071.8	3900.1	5055.1	8.50	5.00	8.14	5.17

(concludes on the following page)

M1	M2	M3	PRIME	DISCOUNT	HOME	T-BILL
1051.8	3863.6	5036.3	8.50	5.00	7.94	5.13
1062.4	3892.1	5070.1	8.50	5.00	7.69	4.92
1063.7	3910.3	5115.3	8.50	5.00	7.50	5.07
1067.5	3944.0	5170.8	8.50	5.00	7.48	5.13
1057.7	3952.4	5197.3	8.50	5.00	7.43	4.97
1054.6	3965.4	5244.3	*	*	*	4.93

CASE 24

SAT SCORES

Many administrators, teachers, and parents perceive a "crisis in education" in the United States. Learned skills and test scores suggest lower achievement by American students both over time and compared to their counterparts around the world. Much of the concern is whether our children will be able to compete in a global economy if they are underprepared to meet the challenges of the coming decades.

Suggested solutions to the crisis in educational achievements are varied. Some argue that instilling a sense of achievement, goal-orientation, and hard work is crucial for motivating our students. Others suggest that the key is to commit more financial resources to the educational system.

Using the SAT scores of high school students as a measure of educational achievement, does it appear that the level of financial resources committed to schools in the U.S. influences educational success? If not, what does influence SAT scores of U.S. students? The data file named SAT contains some information that may assist you in answering these questions. This data set contains 51 observations and 10 variables; these represent SAT scores and other variables for each of the 50 U. S. states and the District of Columbia.

DEFINITIONS

SAT   Average combined verbal/math scores on the Scholastic Assessment Test (ranges from 400 to 1600) for students taking the exam during the 1994–95 school year

An earlier version of these data were compiled by Lisa Radovich, who used the data set in a course project. The data come from U.S. Department of Education, National Center for Education Statistics, Digest of Education Statistics, NCES 96-122, Washington D.C. 1996; and College Entrance Examination Board, "College-Bound Seniors: 1995 Profile of SAT Program Test Takers," 1995.

EXPEND	Current expenditures for public elementary and secondary education for the 1995–96 school year, in billions of dollars
TAKERS	Percentage of 1994–95 school year graduates who took the SAT. This number is calculated using the number of high school graduates in 1995 and the number of 1995 seniors who took the SAT exam.
ENROLL	Total enrollment in public elementary and secondary schools in the United States; fall 1995 figures in thousands of students
TEACHERS	Total number of public elementary and secondary teachers in the United States; fall 1995 figures in thousands of full-time equivalents (FTEs)
RATIO	Student-to-teacher ratio (the average number of students taught per teacher) in the United States, fall 1995. Calculated as ENROLL ÷ TEACHERS
SALARY	Estimated average annual salary of teachers in public elementary and secondary schools for the 1995–96 school year, in thousands of current dollars

STATE

1 = Alabama		27 = Montana	
2 = Alaska		28 = Nebraska	
3 = Arizona		29 = Nevada	
4 = Arkansas		30 = New Hampshire	
5 = California		31 = New Jersey	
6 = Colorado		32 = New Mexico	
7 = Connecticut		33 = New York	
8 = Washington, D.C.		34 = North Carolina	
9 = Delaware		35 = North Dakota	
10 = Florida		36 = Ohio	
11 = Georgia		37 = Oklahoma	
12 = Hawaii		38 = Oregon	
13 = Idaho		39 = Pennsylvania	
14 = Illinois		40 = Rhode Island	
15 = Indiana		41 = South Carolina	
16 = Iowa		42 = South Dakota	
17 = Kansas		43 = Tennessee	
18 = Kentucky		44 = Texas	
19 = Louisiana		45 = Utah	
20 = Maine		46 = Vermont	
21 = Maryland		47 = Virginia	
22 = Massachusetts		48 = Washington	
23 = Michigan		49 = West Virginia	
24 = Minnesota		50 = Wisconsin	
25 = Mississippi		51 = Wyoming	
26 = Missouri			

DATA SET

STATE	SAT	EXPEND	TAKERS	ENROLL	TEACHERS	RATIO	SALARY
1	1029	3.162	8	735.95	44.1	16.7	31.3
2	934	1.173	47	125.26	7.4	16.9	49.6
3	944	3.394	27	766.50	39.6	19.4	32.5

(concludes on the following page)

STATE	SAT	EXPEND	TAKERS	ENROLL	TEACHERS	RATIO	SALARY
4	1005	1.497	6	454.28	29.2	15.6	29.3
5	902	26.842	45	5447.85	225.0	24.2	42.5
6	980	3.338	29	656.28	35.2	18.6	35.4
7	908	4.094	81	514.63	36.0	14.3	50.4
8	897	0.753	68	108.46	6.5	16.7	40.5
9	857	0.503	53	79.80	6.4	12.5	43.7
10	889	11.767	48	2172.79	116.8	18.6	33.3
11	854	6.271	65	1311.13	80.2	16.3	34.3
12	889	0.808	57	186.57	10.2	18.3	35.8
13	979	1.031	15	243.10	12.8	19.0	30.9
14	1048	10.419	13	1927.52	113.0	17.1	41.0
15	882	5.670	58	980.20	55.6	17.6	37.8
16	1099	2.728	5	502.30	32.5	15.5	32.4
17	1060	2.588	9	464.09	30.9	15.0	35.5
18	999	3.683	11	638.63	37.2	17.2	33.0
19	1021	3.499	9	781.14	47.6	16.4	26.8
20	896	1.282	68	219.23	14.6	15.0	32.9
21	909	5.238	64	805.58	46.2	17.4	41.2
22	907	6.302	80	910.02	61.6	14.8	43.8
23	1033	10.140	11	1643.10	80.8	20.3	49.2
24	1085	5.788	9	835.42	48.4	17.3	36.9
25	1036	2.015	4	503.60	28.8	17.5	27.7
26	1045	3.879	9	873.64	57.7	15.1	33.3
27	1009	0.909	21	165.50	10.1	16.4	29.4
28	1050	1.498	9	289.73	20.0	14.5	31.5
29	917	1.247	30	265.04	13.7	19.3	36.2
30	935	1.195	70	190.45	12.2	15.6	35.8
31	898	11.720	70	1197.56	86.9	13.8	47.9
32	1015	2.065	11	328.46	19.3	17.0	29.3
33	892	24.600	74	2830.00	185.9	15.2	48.1
34	865	6.114	60	1165.39	71.8	16.2	30.6
35	1107	0.552	5	119.09	7.5	15.9	27.0
36	975	10.260	23	1838.41	104.2	17.6	37.8
37	1027	2.331	9	616.50	39.3	15.7	28.9
38	947	3.028	51	527.91	26.7	19.8	39.7
39	880	12.916	70	1801.97	104.6	17.2	46.9
40	888	1.098	70	148.08	10.2	14.5	42.2
41	844	2.970	58	637.52	39.2	16.3	31.6
42	1068	0.658	5	144.11	9.3	15.5	26.3
43	1040	3.763	12	880.96	50.7	17.4	33.5
44	893	19.330	47	3740.26	240.2	15.6	32.0
45	1076	1.738	4	473.67	20.1	23.6	30.5
46	901	0.682	68	105.96	7.7	13.8	36.3
47	896	5.957	65	1079.85	73.3	14.7	34.7
48	937	5.585	48	951.70	45.3	21.0	38.0
49	932	1.863	17	306.45	20.9	14.7	32.2
50	1073	5.755	9	869.17	56.2	15.5	38.6
51	1001	0.580	10	99.86	6.7	14.9	31.6

CASE 25

EMERGENCY ADMISSIONS

The emergency department at "General Hospital" is a very busy place, particularly at certain times of the day, certain days of the week, and different months of the year—or so it seems to Cindy, the director of the emergency department at General Hospital. Among other tasks, preparing the weekly staffing schedule is one of Cindy's responsibilities. This is a challenging job, because she needs to make sure that enough nurses, physicians, and other staff members are available to handle incoming emergencies. Yet one of the obvious problems facing all hospitals these days is the budget crunch; Cindy does not want to overstaff the emergency department and pay personnel to stand around with nothing to do.

To understand general trends in admissions to the emergency department, Cindy has collected some historical data from hospital records. In the file named EDADMITS are the data on daily admissions to the emergency department over the period January 1 (a Wednesday) to August 28. (Records are missing for a couple of days, as indicated by an asterisk in the data set.) Cindy has hired you to study the data and provide a forecasting model of the number of admissions to be expected on a daily basis.

Your task is twofold. First, derive a forecasting model and use your model to forecast the number of admissions for August 29, 30, and 31. Explain your model to Cindy in nontechnical terms. Second, prepare a worksheet that Cindy might use to create her own forecasts.

These are real data collected from an emergency department, although we have disguised the name of the institution. These data come from a study project originally prepared by Carolyn Briggs, Christopher Jordan, Christie Martinez, and Joann Schauer.

DATA SET

Read across the row

115	121	118	111	109	103	75	114	103
117	128	76	97	106	78	108	114	124
83	99	122	98	81	128	141	101	121
110	108	106	131	147	104	125	124	121
123	132	145	102	116	114	143	108	140
111	118	117	134	130	104	128	132	122
108	112	112	*	141	129	125	136	134
107	118	137	121	87	112	132	*	111
135	114	111	130	121	99	*	*	105
88	*	*	109	116	112	120	119	118
*	116	134	149	127	110	129	135	137
116	142	128	108	135	126	133	132	164
134	92	127	129	133	148	*	139	*
138	137	129	121	133	135	137	149	130
124	146	151	150	104	127	107	125	133
162	139	96	118	134	149	130	128	146
134	117	143	113	118	150	152	121	143
146	137	121	134	157	128	121	128	133
139	158	156	148	126	157	160	134	143
159	126	138	164	137	132	147	151	104
138	152	127	129	148	149	143	152	126
107	156	138	138	147	147	137	134	140
125	135	102	141	135	136	158	141	146
101	148	143	134	102	127	127	121	179
145	158	141	158	158	128	148	137	127
134	140	136	105	146	135	129	120	129
138	129	146	112	106	105	128		